INTRODUCTION TO ADVANCED MATHEMATICS

Introduction to
Advanced Mathematics
A Guide to Understanding Proofs
Connie M. Campbell
Millsaps College

BROOKS/COLE
CENGAGE Learning

Australia ● Brazil ● Japan ● Korea ● Mexico ● Singapore ● Spain ● United Kingdom ● United States

BROOKS/COLE
CENGAGE Learning™

Introduction to Advanced Mathematics: A Guide to Understanding Proofs, First Edition
Connie M. Campbell

Publisher: Richard Stratton

Senior Sponsoring Editor: Molly Taylor

Assistant Editor: Shaylin Walsh

Media Editor: Andrew Coppola

Art Director, Jill Ort
Sr. Print buyer: Diane Gibbons

Senior Marketing Manager: Jennifer Jones

Marketing Coordinator: Michael Ledesma

MarComm Manager: Mary Anne Payumo

Cover Designer: Wing Ngan
Cover Image: Wing Ngan

Compositor: PreMediaGlobal

For product information and technology assistance, contact us at
Cengage Learning Customer & Sales Support, 1-800-354-9706

For permission to use material from this text or product, submit all requests online at **www.cengage.com/permissions**.
Further permissions questions can be emailed to
permissionrequest@cengage.com.

Library of Congress Control Number: 2010940894

ISBN-13: 978-0-547-16538-7

ISBN-10: 0-547-16538-2

Brooks/Cole
20 Channel Center Street
Boston, MA 02210
USA

Cengage Learning is a leading provider of customized learning solutions with office locations around the globe, including Singapore, the United Kingdom, Australia, Mexico, Brazil and Japan. Locate your local office at **international.cengage.com/region**

Cengage Learning products are represented in Canada by Nelson Education, Ltd.

For your course and learning solutions, visit **www.cengage.com**.

Purchase any of our products at your local college store or at our preferred online store **www.cengagebrain.com**.

Instructors: Please visit **login.cengage.com** and log in to access instructor-specific resources.

Printed in the United States of America
5 6 7 8 9 10 11 22 21 20 19 18

Contents

Preface

This text is designed to help students make the transition from algorithmic to abstract thinking by helping them learn how to read, understand, and construct mathematical proofs. While it was written to be used in a one semester undergraduate course in proof writing, it can also serve as a primer for students who have not been exposed to a dedicated course in proof writing. There are no prerequisites to this material, though a certain level of mathematical maturity is expected. This text is designed not to cover any specific mathematical content, but rather to have an emphasis on helping students learn methods of proof and how to develop their own proofs.

Mathematics is an ever-growing and dynamic field of study. Mathematicians identify patterns, formulate conjectures about those patterns, and then determine the truth or falsity of these conjectures through the method of proof. Simply stated, a proof is a written argument which demonstrates the truth of a statement, using only logic and well established facts. Proofs are the language by which mathematicians add to the current body of knowledge in our discipline, and as such, they are the single most important means of communication in mathematics.

For a number of years now I have taught a course designed to help transition students from more skills-oriented courses such as Calculus and Differential Equations to more theoretical mathematics courses such as Abstract Algebra and Analysis. It has been my experience that providing the right level of difficulty in this class is challenging but also incredibly important. The challenge is that the content should be simple enough that students are instructed in logic and mathematical writing, but it should also be substantial enough that they will be able to transfer their proof writing skills to upper level courses. This text represents what I believe to be the right balance for meeting this challenge.

The text is set up with a three part process: Chapter 1 covers the logic needed to understand proofs, Chapter 2 covers the basic types of mathematical arguments which are used in proof writing, and Chapter 3 is designed to give the student practice in applying their proof writing skills to new concepts and definitions common in upper level undergraduate mathematics courses.

Allow me to describe the content of this learning process in greater detail, as I would introduce the course to a student:

Unlike prior math classes you may have had, the goal of this course is not to find a solution to a problem but rather to take a given solution and construct an argument which explains how and why that specific conclusion may be made. That argument is a proof, and it should make clear the entire reasoning process leading to the solution. This text is designed to help you navigate your way to a given solution and then be able to explain to a third party, in universally accepted mathematical notation, how to get there. The skills needed for such navigating and communicating are presented in each chapter and build upon those of the previous chapter.

Chapter 1 is a study of formal logic as it applies to mathematical thinking. The foci of the sections in this chapter are logic, applying logic, and reading mathematics. We will also cover concepts such as mathematical conjecture, quantifiers, and the concept of mathematical proof. Chapter 2 focuses squarely on reading, writing, analyzing and understanding mathematical proofs. In this unit you will learn how to write a proof (stylistically speaking) and will learn how to apply the logic covered in Chapter 1 to construct several different types of proof arguments.

Once you have completed Chapter 2 you should have the basic skill set needed to understand proof writing and should be ready to strengthen your skills by working with some more complex mathematical definitions (such as definitions which involve the use of multiple quantifiers). In Chapter 3 we will do just this. We will study some specific mathematical topics (set theory, relations, functions, cardinality, etc.) and, in so doing, will examine mathematical definitions, axioms, and mathematical proofs as they are applied in a variety of settings. As you master Chapter 3, you will be gaining the ability to transfer your knowledge of proof writing to a variety of settings, most notably to theory based mathematics courses.

I want to caution you that proof writing is complex, multi-faceted, and requires creativity. It is one thing to understand what a proof is and how a proof should be structured, and it is quite another to be able to develop a proof of a complex statement. Indeed, many mathematicians have spent a good part of their lives trying to develop a particular proof. In the words of W.S. Anglin, "Mathematics is not a careful march down a well-cleared highway, but a journey into a strange wilderness, where the explorers often get lost." When it comes to proof writing, there are no complete maps to show us the way. But, our final proof will be a map for someone coming behind us. Think of this text as an instruction guide to help you use your compass, chart your course, and create your own map.

This text was developed out of my lecture notes and tests. Consequently, each section is presented in the casual, conversational way I would talk with my own students. Additionally, the homework problems are the type of problems I would include on an exam and hence represent the type of skills I would expect my students to master from a section. The text was written in a user friendly language and with a minimalist approach. Most sections are topics I would cover in one 50 minute class period. Due to this minimalist approach I have taken, no section should be skipped or omitted. Rather each section is included for a specific reason and builds upon earlier concepts. I have tried to present the material in a succinct manner with enough examples and activities to empower the students to be able to work the homework on their own. I have also tried to make the shift between sections simple enough that each step is clearly a move onward and upward, but not too steep. I should add that proof writing is not a skill that one can learn by simply watching others or even by just reading a book. To truly develop proof writing skills one must practice writing proofs. Consequently I would discourage the reader from trying to move on to another section before having earnestly tried to work the homework problems in the previous sections. You will do yourself no favors if you omit putting pencil to paper and trying to write your own proofs at each step along the way.

You will note that many of the sections include "In Class Activities" which are examples that I have found useful to work through in class as I covered a section. After I cover the concepts in a section and work through the In Class Activities, I expect my students to be able to perform the homework of the section. As such, the "In Class Activities" are problems a teacher might want to work with a class or, that a person reading through this book on their own should stop and work as they arise. Once a student has been able to understand a section and can work all the in class activities in that section, they should be well prepared to work the problems in the homework.

Teaching a transition course is challenging but incredibly rewarding. There is nothing more enjoyable for me than to watch a student have an "a-ha" moment when they have figured out how to work a problem or to piece together a proof. Constructing proofs is like putting together a puzzle. Some pieces will be obvious and can help you get started, but other pieces may have to be tried over and over at different angles before you find the right fit. Sometimes it seems a piece has been lost and may never be found. But when it is finally complete, the picture is a work of art. In teaching, I strive to help my students find such a sense of joy and accomplishment in their work, and I hope this text extends that joy beyond my own classroom.

I am always interested in ideas for ways to improve teaching this class so if you have suggestions, comments, or even complaints about this text, I hope you will share those with me.

Acknowledgments

I would like to express my appreciation to Millsaps College for providing me with the time, encouragement, and resources which have allowed me the opportunity to work on this project. I would also like to thank my students in Introduction to Advanced Mathematics who, over the years, have encouraged me that I should pursue this project and that it was a worthwhile endeavor. In particular, Jaclyn Fletcher, Amber King, and Tammy Ladner Threadgill were all instrumental in helping to make this project a reality. Additionally, I am grateful to my editor, Dan Seibert, whose keen eye truly helped to improve the book, to Anne Applin who was invaluable in helping me deal with a number of last minute typesetting issues, and to Renee Sappington who helped with the graphics and typesetting, but more importantly, who kept me focused and motivated throughout the process.

Connie M. Campbell, Ph.D.
Professor of Mathematics
Millsaps College
Jackson, MS USA

Chapter 1

Logic

1.1 Introduction and Terminology

The purpose of this course is to help you learn the language of mathematics. Let us begin with a few terms and some common notation for those terms.

A **set** is a collection of distinct objects, and those objects are called **elements** of the set.

Real numbers are numbers which can be marked on a one-dimensional number line. Mathematicians denote the set of real numbers by $\mathbb{R}$.

Natural numbers are those numbers which are used for counting. The common notation mathematicians use to represent the set of natural numbers is $\mathbb{N}$. There is no general agreement among mathematicians as to whether or not this set of numbers should include 0. Hence, some mathematicians consider natural numbers to be the numbers $0, 1, 2, \ldots$ while others consider them to be only $1, 2, \ldots$. For this text, we will not consider 0 to be a natural number but when you read a claim about the natural numbers in other texts, you should check to determine whether or not the author intends for 0 to be included.

Integers are the natural numbers, their negatives, and zero, i.e. the integers are the numbers $\ldots, -2, -1, 0, 1, 2, \ldots$. Mathematicians typically denote the set of integers by $\mathbb{Z}$. This notation comes from the German word zählen which means "count."

Rational numbers are the numbers which can be written as a fraction for which both the numerator and denominator are integers. Note that since a rational number is a real number, the denominator will necessarily be non-zero. Mathematicians denote the set of all rational numbers by $\mathbb{Q}$, which comes from the word quotient.

- .5 is a rational number since it can be written as $\frac{1}{2}$.

- $\sqrt{\frac{9}{4}}$ is a rational number since it can be written as $\frac{3}{2}$.

- $\sqrt{3}$ is not a rational number since it cannot be written as an integer divided by an integer (we'll prove this fact later).

You may recall from prior studies that any number which, when written using its decimal form, either terminates or repeats indefinitely is a rational number.

- 1.345 can be written as $\frac{1345}{1000}$.

- 1.134343434..., typically written as $1.1\overline{34}$, can be written as $\frac{1123}{990}$.

If you don't remember how to rewrite a repeating decimal as the quotient of two integers, consider the following argument which can be generalized to show that any real number for which a decimal portion repeats indefinitely can be written as an integer divided by an integer:

> Let $n = 1.1\overline{34}$. Note that if we multiply n by 10 we will isolate the repeated portion of the number to the decimal position. Now we can also multiply n by 1000 to yield a number with the same decimal portion. That is $1000n = 1134.\overline{34}$ and $10n = 11.\overline{34}$. Note then that we can combine these two equations as follows to eliminate the repeated decimal portion.
>
> $$\begin{aligned} 1000n &= 1134.\overline{34} \\ -\ 10n &= \ \ \ 11.\overline{34} \\ \hline 990n &= 1123 \end{aligned}$$
>
> From this we can see that $n = \frac{1123}{990}$. As we have shown that n can be written as the quotient of two integers, we know that n is a rational number.

The argument above should give you a general sense of a detailed mathematical explanation, and the construction and writing of such explanations (or proofs) will be the main focus of this course. But for now we continue with some terminology.

Irrational numbers are real numbers which are not rational. That is, irrational numbers are real numbers which cannot be written as an integer divided by an integer.

Positive numbers are real numbers which are greater than 0. Mathematicians typically denote the set of positive real numbers by $\mathbb{R}^+$ and the set of positive integers by $\mathbb{Z}^+$.

Negative numbers are real numbers which are less than 0. Mathematicians typically denote the set of negative real numbers by $\mathbb{R}^-$ and the set of negative integers by $\mathbb{Z}^-$.

Non-negative numbers are real numbers which are not negative and hence this is a way to say all numbers greater than or equal to zero. Mathematicians typically denote the set of non-negative real numbers by $\mathbb{R}^*$ and the set of non-negative integers by $\mathbb{Z}^*$. Similarly, **non-positive numbers** include all real numbers less than or equal to zero.

So, a set is a collection of distinct elements and $\mathbb{R}, \mathbb{Z}, \mathbb{Q}$ and $\mathbb{N}$ are all examples of sets. There are several appropriate ways to write a set. One way is to assign the set a name (such as $\mathbb{Z}$ or $\mathbb{R}$ or A). Another way is to list the set's elements inside braces. For example $\mathbb{Z} = \{..., -3, -2, -1, 0, 1, 2, 3, ...\}$. Either of these notations is appropriate. If we write $A = \{1, 2, 3\}$ then we are saying that A is a set whose elements are the numbers $1, 2,$ and 3. So we could say that 1 is an element of the set A but 5 is not an element of the set A. The braces are important as $A = 1, 2, 3$ has no meaning. It actually reads that $A = 1$ and then the 2 followed by the 3 would be considered improper grammar.

Another shorthand notation with which you should be familiar is the symbol "$\in$". This symbol means "is an element of." So, using the set A above, we see that $3 \in A$. This literally reads "3 is an element of A." The shorthand notation mathematicians use to state that something is not an element of a set is $\notin$. So, for our set above, $5 \notin A$. The strike over key indicates "not" so it simply reads "5 is not an element of A."

The use of a vertical bar inside of braces is shorthand for "such that" or "where." For example, instead of writing $A = \{1, 2, 3\}$, we could write $A = \{x$ is a natural number $|x < 4\}$ or $\{x \,|x$ is a natural number and $x < 4\}$ or $A = \{x \in \mathbb{N} | x < 4\}$. The first of these three alternatives would read "A is the set of all natural numbers x, where x is less than 4 "; the second would read "A is the set of all x such that x is a natural number less than 4 "; and the last would read "A is the set of all natural numbers which are less than 4." Another shorthand notation for the phrase "such that", and one which is often used in mathematical writings, is the symbol "$\ni$".

One last set notation that we mention here is interval notation. You have used this notation in algebra or calculus to represent certain sets of real numbers. For example $[3, 5)$ represents the set of all real numbers which are greater than or equal to 3, and less than 5. Using the braces notation we could write $[3, 5) = \{x \in \mathbb{R} \mid 3 \leq x < 5\}$. Either of these notations represent the same set of numbers but the two sets $[3, 5]$ and $\{3, 4, 5\}$ represent different sets of real numbers as the latter does not include many elements which are in the first (e.g. 3.7).

Homework 1.1:

Determine if each of the following statements are true or false and justify the reason for your answer:

1. If $x \in \mathbb{R}$, then $x \in \mathbb{Z}$.

2. If $x \notin \mathbb{R}$, then $x \in \mathbb{Z}$.

3. If $x \in \mathbb{N}$, then $x \in [1, \infty)$.

4. If $x \in \mathbb{R}$, then $x \notin \mathbb{Z}$.

5. If $x \notin \mathbb{R}$, then $x \notin \mathbb{Z}$.

Rewrite the following repeating decimals in the form of an integer divided by an integer:

6. $1.\overline{2}$

7. $.23\overline{45}$

1.2   Statements and Truth Tables

A **statement** is a sentence or group of sentences which, taken as a whole, is either true or false. Now you might think that everything must be either true or false but this is not the case.

A sentence along the lines of "Chocolate ice cream is the best" is neither true nor false as a whole. It is an opinion that might be stated by someone. If that someone clarifies that the sentence applies to them then it would indeed be a statement since, for them, it would either have to be true or false. It could also be revised to read "Chocolate ice cream is everybody's favorite dessert" or "Dr. Fletcher thinks that chocolate ice cream is the best." These sentences are indeed statements.

Similarly, the sentence "$x - 1 = (x + 2)(x + 3)$" is not a statement since there is no clarification of what is meant by "x". The sentence is ambiguous even though you might want to assume that x represents a real number (just as you might have assumed that the original sentence about ice cream was referring only to the person who wrote the sentence).

If there are variables involved, for the sentence to be a statement, those variables must be defined. Hence, for the sentence referred to above to be either true or false, x must have some sort of meaning.

Some good examples would be "If x is a real number, then $x - 1 = (x+2)(x+3)$" or "For any real number x, $x - 1 = (x + 2)(x + 3)$" or "Let $x \in \mathbb{R}$. $x - 1 = (x + 2)(x + 3)$." Each of these forms a statement (in fact you will soon learn that they all make the same claim).

The point is that any and all variables must be defined for a sentence, or group of sentences, to be a statement. When the variables are well defined, we can evaluate whether the statement is true or false. In the case of the statement above: "For any real number x, $x - 1 = (x + 2)(x + 3)$" this is a false statement since it claims that any real number will satisfy the equation and this is not true for 3. That is $x = 3$ is a real number but $3 - 1 \neq (3 + 2)(3 + 3)$.

In general, by the very fact that a sentence or group of sentences is declared to be a statement, you definitively know that the statement will be either true or false.

Another type of sentence which fails to be a statement is a paradox. An example of such would be "This sentence is false." The reason that sentence cannot be true is that, if it is, it certainly can't say what it does (that it is false). Similarly, if it is false, then since it says it's false, it's true. Neither of these options makes sense; hence, the sentence cannot be true or false; it is not a statement.

In mathematics we often work with statements, and we try to determine the truth value of statements. If we discover that a statement is true, we can write a mathematical argument explaining this to others (such a piece of writing is called a proof). If we discover that a statement is false, we can write a mathematical argument explaining why the statement is false (often called a disproof).

In this course, we will only deal with mathematical statements; hence, you can be confident that if the statement is not true, it must be false. Mathematicians continually try to evaluate the truth value of mathematical statements. Consequently we will be studying the language of mathematics as it applies to evaluating statements about mathematics and determining the truth value of those statements.

A written argument which demonstrates the truth of a statement, using only logic and well established facts, is called a proof. The primary goal of this course is for you to learn the process of proof writing so that you might have the skill set necessary to continue on your journey in mathematics.

In this course, we will begin by studying the basic rules of logic as well as some common types of proof arguments. You will then learn to write some proofs on an elementary level and, as you learn more mathematics, your tool set will increase, and you will learn to prove more complicated mathematical statements.

Consider the following statement:$\sqrt{2}$ is an integer. The negation of this statement would be "It is not true that $\sqrt{2}$ is an integer" or more succinctly "$\sqrt{2}$ is not an integer." The **negation** of a statement is the statement (regarding the same content) which has the opposite truth value of the original statement. If the original statement is true, then it's negation will be a statement (regarding the same content) which is false.

There are other ways that statements can be modified to produce new statements. For example, if you have two statements and you combine them with the word *and*, the result will be a statement. This new statement's truth value depends on the two statements with which you began. For example, if you have a statement p which is true and a statement q which is false, then the statement "p and q" would be a false statement. A statement of the form "p and q" is often called a conjunction.

Another way to combine two statements to yield another statement is with the word *or*. Again, the truth value of the resulting statement will depend on the truth value of the two original statements. If p is a true statement and q is a false statement, then the statement "p or q" would be true.

There are actually two ways that the word *or* gets used. Sometimes when we use *or* we are asking the person to choose one of two or more options. For example, a waiter asks, "Will you have the chicken marsala or the spinach lasagna?" In this usage he is asking for you to choose one, or the other, but not both. This type of use of or is called an "exclusive or."

A second type of use of the word *or* is an "inclusive or." Suppose a student asks, "Have you taken Mrs. Dance or Ms. Sullivan for a math class?" They are interested to know if you have taken a course under either (or both) of these professors. This use of *or* is called an "inclusive or."

In mathematics any use of the word *or* is assumed to be an "inclusive or" unless it has been clearly stated otherwise. Consequently, if p is a true statement and q is a true statement, then the statement "p or q" is a true statement. A statement of the form "p or q" is often called a disjunction.

Another way to combine two statements to yield another statement is with the word *implies*. If p is a statement and q is a statement, then "p implies q" or equivalently "If p, then q" is also a statement. This type of statement is called a conditional statement. In a statement of this form, p is called the hypothesis and q the conclusion.

Regarding conditional statements, whenever the hypothesis is true, in order for the entire statement to be true, the conclusion must also be true. Also, whenever the hypothesis is false, the statement is true independent of whether or not the conclusion is true. To help you understand this last point think about it as a legal issue. Suppose your teacher tells you, "If you come to class then you will pass," but you fail the class. You decide to sue the teacher and as part of the litigation you explain what your teacher said. The first question you would be asked is, "Did you come to class?" If you did, then the teacher has clearly lied, and you will win the case. In the language of our statement, a lie means that the hypothesis of the statement was true, but the conclusion did not follow. However, if you did not go to class then you cannot claim the teacher lied. You did not come to class, and so the teacher's statement is not false. Since the teacher didn't tell a lie, she told the truth (remember statements have to be either true or false so if it's not false, it's true).

"If, then" statements are extremely important in mathematics and consequently we will elaborate more on this type of statement later. But we end this section with one more connective (which is really just a combination of some of those we have already studied) and that is "if and only if." Statements connected by "if and only if" are also statements. Another way to phrase "p if and only if q" is to say "p if q, and p only if q." That is, "p if and only if q" is a conjunction of two implication statements with the first implication being "if q, then p", and the second being "if p, then q." As such, if and only if statements are often referred to as **biconditional** statements. Statements of this form are so common in mathematics that there is a standard shorthand for "if and only if", and that is "iff." So, if you ever see "iff" written in a mathematics text, it is not a spelling error of the word "if" but rather a shorthand notation for "if and only if." You should feel free to use this shorthand in any of your own mathematical writings.

For the purpose of this chapter, we will also define some shorthand symbolic notation for the methods of combining statements we have just considered. These symbols are often called logical connectives or logical operators.

$\neg p$	symbolizes	not p
$p \wedge q$	symbolizes	p and q
$p \vee q$	symbolizes	p or q
$p \rightarrow q$	symbolizes	If p, then q.
$p \leftrightarrow q$	symbolizes	p iff q

Although this notation is handy, several of these symbols have other meanings in different settings, so you are encouraged to only use these shorthand notations if the context is quite clear as to their meaning. Consequently, while these notations are commonly used in mathematics classes, these are not symbols that should be used in a written proof. Rather, in a proof the actual words *not, and, or, implies,* and/or *iff* should be used.

A **truth table** is a diagram which details all the possible outcomes for a given statement. For example, if the statement only involves one basic component, such as "4 is an even integer", then as there are only two possible outcomes. However, a statement may involve the combination of several statements, say for example "If x is an even integer, then x is a multiple of 2 and $x^2 - 1$ is an odd integer." While this new statement is still either true or false as a whole, its truth outcome depends on the truth value of each of the component statements. To see this, note that the statement above is of the form "$p \rightarrow (q \wedge r)$" where each of p, q, and r are considered simple statements. Here p represents "x is an even integer", q represents "x is a multiple of 2", and r represents "$x^2 - 1$ is an odd integer." We know that our original statement is either true or false and a truth table will help us to see what conditions on p, q, and r will yield our final statement being true, and/or what conditions on p, q, and r will yield a false statement.

Now we present the diagram form. If your statement is simply of the form p then a truth table for this statement would look like the following.

p
T
F

If your statement was of the form "p and q" then your truth table should look something like this next table.

p	q	$p \wedge q$
T	T	T
T	F	F
F	T	F
F	F	F

In the last table, we listed all the possible combinations for the truth values of each of the components of our statement and then, in the final column, included the resulting truth value of our actual statement. Reading this chart tells us that the only way for a statement of the form "p and q" to be true is if both p and q are true — all other combinations yield a false result. This means that in order to prove a statement such as "continuous functions are differentiable and step functions are monotonic" we would need to provide a written explanation that explained why it must be true that continuous functions are differentiable, and also that step functions are monotonic. If we wanted to disprove this statement then we could do it in a number of ways. We could simply explain to our reader that "continuous functions are differentiable" is a false statement. That would do it since if p is false, then "p and q" must be a false statement. What is another way that we could disprove this statement?

In Class Activity: Write out a truth table for "p or q" as well as a truth table for "$p \rightarrow q$."

Since the truth table lists every possible truth value for all the components of a statement, the number of rows in the truth column depends on how many simple components the statement has. For example, a truth table for a simple statement of the form p would just have two rows as p must either be true of false. A truth table involving two components (such as "p and q") has four rows since there are that many combinations for the truth values of the two components. Recall our earlier statement, "If x is an even integer, then x is a multiple of 2 and $x^2 - 1$ is an odd integer." Suppose that we wanted to determine under precisely which conditions this statement, which is of the form "$p \rightarrow (q \wedge r)$", is true. We could discern this by producing a truth table:

p	q	r	$q \wedge r$	$p \rightarrow (q \wedge r)$
T	T	T	T	T
T	T	F	F	F
T	F	T	F	F
F	T	T	T	T
T	F	F	F	F
F	F	T	F	T
F	T	F	F	T
F	F	F	F	T

From the truth table it is clear that a statement of this form will be true unless p is true and r or q (or both) is false.

So truth tables are a tool which allow us to easily see what conditions must be met for a statement to be true or false. This comes in handy particularly if you have a complex statement (one involving several components connected by *and, implies, or,* or *not*) and you are having trouble understanding exactly what it would mean for the statement to be true or false.

Recall that the number of possible combinations (and so the number of rows of the truth table) depends on the number of simple components in the statement. When our final statement involved only a simple component p, then there were only two options regarding all combinations of the truth values of the components, but when there were three components $[p \rightarrow (q \wedge r)]$, there where eight such combinations. In general, if your statement has n simple components, your truth table will have 2^n rows.

Another important thing to note about truth tables is that the only required columns are those of the individual components (p, q, r, etc.) and the last column should contain the statement you are evaluating. Any columns added between these are simply a way to keep track of the components of the final statement so as to keep yourself from making errors.

Finally, the order of the rows in a truth table is not significant. That is, if you were working with a statement of the form "p or q", it doesn't matter how you set up your truth table:

p	q	$p \vee q$
T	T	T
T	F	T
F	T	T
F	F	F

p	q	$p \vee q$
T	F	T
F	F	F
F	T	T
T	T	T

You could also set up your rows in any other order. Both of the truth tables show that the only time "$p \vee q$" is false is when both p and q are false, and that is the point of a truth table.

We close this section with a note about the order of precedence of logical connectives. In general, the order of operations for operators is negation, conjunction/disjunction, implication. For example, a statement of the form $p \vee q \rightarrow \neg\, r \wedge m$ would be considered as $(p \vee q) \rightarrow (\neg\,(r) \wedge m)$. If you wrote $p \vee q \rightarrow \neg\, r \wedge m$, when you intended $p \vee (q \rightarrow \neg\, r) \wedge m$, then you would need to include the parenthesis (or in your writing you'd use commas), to specify this alternate order to your reader.

Homework 1.2:

1. Construct a truth table for $p \rightarrow \neg p$.

2. Construct a truth table for $p \wedge (q \rightarrow p)$.

3. Construct a truth table for $p \wedge q \rightarrow p$.

4. Use the truth table you constructed for problem 1 above to determine under what conditions (if any) that statement would be true.

5. Use the truth table you constructed for problem 2 above to determine under what conditions (if any) that statement would be false.

1.3   Logical Equivalence and Logical Deductions

Two statements are said to be **logically equivalent** if they contain the same simple components and, no matter what the truth values of the simple components of these two statements are, the two statements have the same truth value.

For example, consider the following two statements:

1. If $f(x)$ is differentiable at x, then $f(x)$ is continuous at x.
2. If $f(x)$ is not continuous at x, then $f(x)$ is not differentiable at x.

These two statements involve the same simple components. If we denote "$f(x)$ is differentiable at x" as p and "$f(x)$ is continuous at x" as q, then we can rewrite these two statements as follows:

1. $p \to q$.
2. $\neg q \to \neg p$.

We construct a truth table which includes each of these — the third column is our first statement and the last column is our second statement.

p	q	$p \to q$	$\neg q$	$\neg p$	$\neg q \to \neg p$
T	T	**T**	F	F	**T**
T	F	**F**	T	F	**F**
F	T	**T**	F	T	**T**
F	F	**T**	T	T	**T**

Note that the column entries for "$p \to q$"and "$\neg q \to \neg p$" agree in each and every case. Hence, these two statements are logically equivalent. If we know that one of these two statements is true then we know for certain that the other must also be true. Alternatively if we know that one of these two statements is false, we also know definitively that the other is false as well.

As another example consider the following statements:

1. I am not going to work and I am not going out to eat.
2. If I am going to work, then I am going out to eat.

As above, let us abbreviate the simple components of these statements. Let p represent "I am not going to work" and q represent "I am not going out to eat." Using our new notation we would like to know if "$p \wedge q$" is logically equivalent to "$\neg q \to \neg p$." We begin the process by constructing a truth table for each:

p	q	$p \wedge q$	$\neg q$	$\neg p$	$\neg q \to \neg p$
T	T	T	F	F	T
T	F	F	T	F	**F**
F	T	**F**	F	T	**T**
F	F	F	T	T	

We can stop at this point because we see from the bold entries that if p is true and q is false, one of the original statements will be true while the other will be false. Hence, they are not logically equivalent. Knowing that one of these statements is true will not assure you that the other is also true.

Logical equivalences are very important as you will see later in the text. Here are some of the most commonly used logical equivalences (so commonly used that you will want to commit these to memory).

p	is logically equivalent to	$\neg(\neg p)$
$p \rightarrow q$	is logically equivalent to	$\neg q \rightarrow \neg p$
$\neg(p \vee q)$	is logically equivalent to	$\neg q \wedge \neg p$
$\neg(p \wedge q)$	is logically equivalent to	$\neg p \vee \neg q$
$\neg(p \rightarrow q)$	is logically equivalent to	$p \wedge \neg q$
$p \rightarrow q$	is logically equivalent to	$\neg p \vee q$
$p \leftrightarrow q$	is logically equivalent to	$\neg p \leftrightarrow \neg q$
$(p \vee q) \rightarrow r$	is logically equivalent to	$(p \rightarrow r) \wedge (q \rightarrow r)$

Now that we've discussed statements and what makes those statements true or false we return to implications. "If, then" type statements occur frequently in mathematics and warrant a more thorough study.

We saw that a statement of the form "if p, then q" is false only in the case that p is true and q is false. What if you have an implication statement that you know is true and you know something about one of the components? Could you use that information to determine the truth value of the other component? For example, suppose you know that "If x is a multiple of 6, then x is even" is a true statement and you know that "x is not even" is also true. Could you conclude from this information that x is not a multiple of 6?

Referring back to the truth tables this would be a situation where the statement "$p \rightarrow q$" is true and q is false. In this case you would know that p must be false and so, in context, this means that you could indeed definitively conclude that x is not a multiple of 6.

Taking this a bit further, suppose that you know that "If x is a multiple of 6, then x is even" is a true statement and you know that "x is even." Could you conclude here that x is a multiple of 6? Again, referring to the truth table language, we know that statement "$p \rightarrow q$" is true and q is true. In this case we cannot make a conclusion about p. Here p may be true or false since either of these options, along with q being true yields that "$p \rightarrow q$" is a true statement.

This process of combining several statements together to see if a conclusion can be made is called making a logical deduction.

In Class Activity 1: If you know that "$p \rightarrow q$" is a true statement and you know that p is a true statement, can you conclude that q is a true statement? Why or why not?

We now present an interesting application of the process of determining whether or not two statements are logically equivalent.

Consider Chvátal's Theorem (an important result in the area of graph theory) as it is stated by Buckly and Lewinter[1]:

> Let G be a graph with n vertices ($n > 3$) with degrees $d_1 \leq d_2 \leq ... \leq d_n$. If $d_i \leq i < \frac{n}{2}$ implies that $d_{n-i} \geq n - i$, then G is hamiltonian.

Now consider Chvátal's Theorem as stated by Douglas West[2]:

> Let G be a graph with vertex degrees $d_1 \leq d_2 \leq ... \leq d_n$, where $n > 3$. If $i < \frac{n}{2}$ implies that $d_i > i$ or $d_{n-i} \geq n - i$, then G is hamiltonian.

It is not obvious that these two statements make the same claim but let us start the thought process of verifying that they are indeed logically equivalent. Note that the first sentence of both forms of the theorem agree but the second sentence in each is stated differently. If we rewrite Buckly and Lewinter's version to "if (p and q) implies r, then s", West's second sentence would be "if q implies (not p or r), then s."

Before we construct a truth table with four component pieces to see if these are indeed logically equivalent, note that they are both implication statements with the same conclusion. Hence, we need only confirm that the hypotheses are equivalent. That is that "(p and q) implies r" is logically equivalent to "q implies (not p or r)."

> **In Class Activity 2:** Construct a truth table to show that the two different forms of Chvátal's Theorem provided above are indeed logically equivalent.

Before we end this section, we would like to point out that there are a number of ways to write an implication statement. For example, the statement "For any integer n, n^2 is even" has the same meaning as "If n is an integer, then n^2 is even." Also the statement "A continuous function must be integrable" has the same meaning mathematically as "If f is a continuous function, then f is integrable."

---

[1] Fred Buckley and Marty Lewinter, *A Friendly Introduction to Graph Theory* (Upper Saddle River, New Jersey: Prentice Hall, 2003), 159.

[2] Douglas West, *Introduction to Graph Theory* (Upper Saddle River, New Jersey: Prentice Hall, 2001), 290.

Until now we have only written "p implies q" or "if p, then q" but you should be acquainted with a number of other forms of this type of statement as well, all of which are commonly used in mathematical writing. Each of the following is a way to represent the same statement, a statement that we would write symbolically as "$p \rightarrow q$."

p implies q.
If p, then q.
q if p.
For q to occur, it is enough that p occurs.
p is sufficient for q.
For q, p is sufficient.
For p to occur, q must occur.
p does not occur without q.
p only if q.
q is necessary for p.
For p, q is necessary.
q unless not p.
p cannot occur without q occurring.
q whenever p.
q provided that p.

In Class Activity 3: Write each of the following statements in the form "If_, then_."

1. A set that is compact is closed and bounded.

2. $f(x) = f(y)$ implies $x = y$.

3. For H to be closed under subtraction, it is sufficient that H is closed under addition and contains inverses.

4. For G to be a group, it is necessary that G contain inverses.

5. f is a function only if f is well-defined.

6. $|f(x) - f(a)| < \varepsilon$ whenever $|x - a| < \delta$.

7. All integers are real numbers.

Homework 1.3:

1. Show that $\neg(p \vee q)$ is logically equivalent to $\neg q \wedge \neg p$.

2. Show that $\neg(p \rightarrow q)$ is logically equivalent to $p \wedge \neg q$.

3. Show that $(p \vee q) \rightarrow r$ is logically equivalent to $(p \rightarrow r) \wedge (q \rightarrow r)$.

4. Explain why $p \rightarrow q$ is not logically equivalent to $q \rightarrow p$.

For problems 5 through 7, determine whether or not the two statements given are logically equivalent and explain your reasoning.

5. $p \rightarrow q$.

   $p \vee \neg q$.

6. If $x \notin S$, then $xy \notin T$.

   $xy \in T$ or $x \notin S$.

7. $p \leftrightarrow q$.

   $(p \wedge q) \vee \neg(p \vee q)$.

8. If you know that "$p \rightarrow q$" is a true statement and you know that q is a true statement, can you conclude that p is a true statement? Why or why not?

9. If you know that "$p \rightarrow q$" is a true statement and you know that p is a false statement, can you conclude that q is a false statement? Why or why not?

   If you know that "$p \rightarrow q$" is a true statement and you know that q is a false statement, can you conclude that p is a false statement? Why or why not?

10. Suppose you know that "If G is hamiltonian, then G is connected" is a true statement, and you know that "G is connected" is also true. Can you conclude that G is hamiltonian?

11. Suppose you know that "If G is not abelian, then G is not cyclic" is a true statement, and you know that "G is abelian" is also true. Can you conclude that G is cyclic?

12. Using the four true statements below, can you logically conclude that "x is even iff x^2 is even" is a true statement? Why or why not? x is an integer. Integers are either even or odd, but not both. If x is even, then x^2 is even. If x is odd, then x^2 is odd.

13. If you know that "$p \rightarrow q$" is a true statement and you know that p is a true statement, can you conclude that q is a true statement? Why or why not?

14. Under what conditions (on p and/or q) would a statement of the form "$q \lor \neg(q \lor \neg p)$" be true? If you aren't sure just from looking at the statement, you may want to use a truth table.

Rewrite each of the following statements in the form of "If _ , then _."

15. A continuous function must be integrable.

16. A sequence which is not bounded cannot have a limit.

17. Being non-negative is a necessary condition for a number to have a square root.

18. A one-to-one function has an inverse.

19. A sequence is increasing provided that for all i, $a_i < a_{i+1}$.

20. $ab^{-1} \in H$ only if $aH = bH$.

21. For all $x \in \mathbb{Z}$, $3x^2 - 2x + 1 \in \mathbb{Z}$.

22. For a tree to be good, it is necessary that the distance sum is even.

23. Solving problem 8 is sufficient for solving problem 7.

24. A function is continuous whenever it is differentiable.

25. Reading a good book is sufficient to keep me happy all day.

26. I watch tv whenever I have time.

27. To get into Mrs. Dance's section, you have to sign up during pre-registration.

28. For me to do well in a mathematics class, I have to study.

29. I'm going to fail this test unless somebody shows me how to work number 3.

1.4   The Contrapositive and Converse of an Implication Statement, and the Negation

We introduced the concept of the negation of a statement in Section 1.2 and then saw in Section 1.3 that the negation of a statement of the form $p \wedge q$ would be $\neg p \vee \neg q$. Similarly, the negation of an "or" statement yields an "and" statement. In this section we elaborate on this concept of negation as well as introduce the concepts of the contrapositive and the converse of a conditional statement. We begin by recalling that, as we saw in Section 1.3, $p \to q$ is logically equivalent to $\neg q \to \neg p$, $\neg(p \to q)$ is logically equivalent to $p \wedge \neg q$, and $p \to q$ is not logically equivalent to $q \to p$. These are three very important relationships and which deserve some elaboration.

The **contrapositive** of a statement of the form "If p, then q" is "If not q, then not p." As a statement and its contrapositive are logically equivalent, if you know one of these is true the other must be true as well (and vice versa).

The **negation** of a statement of the form "If p, then q" is "p and not q." As a statement and its negation have the opposite truth value, knowing that one of these statements is true allows you to conclude that the other (its negation) must be false (and vice versa). We will discuss the negation of an implication statement further in the next section since it is often easier to think of the negating an implication if you write the implication in terms of a quantifier. For a feature of coming attractions, another way to think about negating "If p, then q" is "There is a p such that $\neg q$."

The **converse** of a statement of the form "If p, then q" is "If q, then p." As a statement and its converse are not logically equivalent, knowing that one of these statements is true gives you no information about the truth value of the other statement.

As an example consider the following two implication statements:

1. If x is an integer, then x is not a real number.
2. If x is an integer, then $2x$ is an integer.

Note that the first of these statements is false and the second is true. Is this clear? Regarding the first statement, integers are real numbers. Hence, the first statement is not true. However, the second statement is true because if you multiply any integer by 2 the result will still be an integer.

Now, if we form the converse of each of these we obtain the following:

1. If x is not a real number, then x is an integer.
2. If $2x$ is an integer, then x is an integer.

Note that the first of these statements is false (as was its converse), and the second is also false (as opposed to its converse which was true). The point is that a statement and its converse are not logically equivalent and so knowing that a statement is true (or false) gives you no information as to whether or not its converse is true (or false).

To summarize some of the key concepts:

The negation of	p	is	$\neg p$
The negation of	$p \vee q$	is	$\neg p \wedge \neg q$
The negation of	$p \wedge q$	is	$\neg p \vee \neg q$
The negation of	$p \rightarrow q$	is	$p \wedge \neg q$
The negation of	$p \leftrightarrow q$	is	$(p \wedge \neg q) \vee (q \wedge \neg p)$
The converse of	$p \rightarrow q$	is	$q \rightarrow p$
The contrapositive of	$p \rightarrow q$	is	$\neg q \rightarrow \neg p$

In Class Activity: Write the contrapositive, converse, and negation of each of the following statements:

1. If x is even then x^2 is even.

2. If x is a real number then $x > 0$ or $x \leq 0$.

3. If x is a multiple of 6, then x is divisible by 2 and x is divisible
   by 3.

Homework 1.4:

For problems 1 through 9, write the contrapositive, the negation, and the converse of the given statement:

1. $p \rightarrow q$.

2. $p \wedge q \rightarrow r$.

3. If x is even, then $x^2 + x$ is even.

4. If $G_1 \approx G_2$ and G_1 is cyclic, then G_2 is abelian.

5. If f is not differentiable at $x = a$, then f is not continuous at $x = a$.

6. If x is an even integer or $x > 17$, then x is a multiple of 4 and $x > 5$.

7. If $ab = 0$, then $a = 0$ or $b = 0$.

8. If x is divisible by 4 or 6, then x is not prime.

9. If $2^m - 1$ is prime then m is prime.

10. Write the negation of p iff q.

11. An integer x is divisible by 6 iff x is divisible by 2 and x is divisible by 3. State what it would mean if an integer x is *not* divisible by 6. That is, complete the following: An integer x is not divisible by 6 iff _____.
    Helpful hint: Remember that $p \leftrightarrow q$ is logically equivalent to $\neg p \leftrightarrow \neg q$.

12. A function $f(x)$ is said to be 1-1 iff $f(a) = f(b)$ implies $a = b$. Explain what it means to say that a function $f(x)$ is *not* 1-1.

13. An integer x is said to be **terrific** iff x divides ab implies that x divides a or x divides b. Explain what it means to say that an integer x is *not* terrific.

1.5 Quantifiers

Consider the following three statements:

- There is a real number x such that $x + 1 = 8$.
- There is exactly one real number x which is a solution to the equation $x + 1 = 8$.
- For all real numbers x, $x + 1$ is a real number.

Each of these statements refers to real numbers and each has a phrase that tells you the quantity of real numbers about which the statement is making a claim, be it at least one, exactly one, or every one. For the first statement to be true, we need to be sure that there is a real number that will make $x + 1 = 8$. For the second statement to be true, we need to be sure that there is one such real number, and that it is the only such real number. That is, there is one such x and there is not another real number which has this same property. For the third statement to be true, we need to know that for all real numbers, adding one yields a real number.

Such descriptives are called quantifiers and the three quantifiers we used above are very common in mathematics. So common in fact that, you guessed it, mathematicians have a shorthand notation to abbreviate these phrases. $\exists$, $\exists!$, and $\forall$ are the notations that are generally accepted as shorthand for the quantifiers used above. The first two of these, $\exists$, and $\exists!$ are called existential quantifiers, while the last is called a universal quantifier.

$\forall$ represents "for every" or "for all" or "for each,"

$\exists$ means "there is" or "there exists" or "for some," and

$\exists!$ means "there is one and only one" or "there is exactly one" or "there is a unique" or "there is a distinct."

So, to the mathematician, "$\forall x \in \mathbb{Z}$, $\exists y \in \mathbb{Z}$ such that $x + y = 0$" is a perfectly well written statement. In English it states "For every integer x, there is an integer y such that $x + y = 0$." Stated another way, "If x is an integer, then there is an integer y such that $x + y = 0$."

Another phrase, which occurs with such frequency that there is a mathematical shorthand notation for it, is the phrase "such that." The accepted mathematical notation for "such that" is "$\ni$". In particular this phrase is used along with an existence quantifier as "such that" is another way to say "which has the property that" When we write a statement using "for every" or "there exists," the first of these denotes that a property is going to follow and that this property holds for all. The second exhorts that there is a time when the following property holds. That is, we would write "For every real number x, x^2 is non-negative" and we would write "There exists a real number x, such that x^2 is non-negative." Note that if in this latter sentence we omitted the "such that" and simply wrote "There exists a real number x, x^2 is non-negative" our sentence would not be grammatically correct.

In Class Activity 1:   Write the following statements using quantifiers:

1. If $x, y \in \mathbb{R}$, then $x + y \in \mathbb{R}$.

2. There is an $\varepsilon > 0$ such that $N_\varepsilon(x) \subset A$.

3. For every real number $\varepsilon > 0$, there is a real number $\delta > 0$, such that if $0 < |x - a| < \delta$, then $|f(x) - L| < \varepsilon$.

4. For any integer x, you can find unique integers q and r such that $x = 3q + r$ with $0 \leq r < 3$.

5. $a_i > 3$, for some $i \in \mathbb{N}$.

6. $a_i > 3$, for all $i \in \mathbb{N}$.

In Class Activity 2: Determine if each of the following statements are true or false and explain your reasoning:

1. $\exists x \in \mathbb{Z} \ni \frac{x}{10} \in \mathbb{Z}$.

2. $\exists! x \in \mathbb{Z} \ni \frac{x}{10} \in \mathbb{Z}$.

3. $\exists! x \in \mathbb{Z} \ni x + 1 = 6$.

4. $\forall x \in \mathbb{N}, \frac{1}{x} \in \mathbb{R}$

5. $\forall x \in \mathbb{Z}, \frac{1}{x} \in \mathbb{R}$

Now we consider some more complex statements involving quantifiers:

$$\forall x \in \mathbb{Z}, \forall y \in \mathbb{Z}, x + y \in \mathbb{Z}.$$
$$\exists x \in \mathbb{Z} \ni \forall y \in \mathbb{Z}, x + y = y = y + x.$$

The first statement above involves two quantifiers (two universal quantifiers) and it states that if you take any two integers x and y and add them, the result will be an integer. Another way this statement might be written is $\forall x, y \in \mathbb{Z}, x + y \in \mathbb{Z}$.

The second statement above also involves two quantifiers (an existence quantifier and a universal quantifier) and it states that there is an integer x which has the property for any integer y, $x + y = y = y + x$. This is a true statement since 0 is such an integer. That is, if we let $x = 0$, we see that x is an integer with the desired property that for any integer y, $x + y = y = y + x$.

Next consider the following: $\forall x \in \mathbb{Z}, \exists y \in \mathbb{Z} \ni x + y = 0$. This statement is also true since it states that for any integer x, you can find an integer y which, when added to x, yields a sum of 0.

Statements involving multiple quantifiers are quite frequent in mathematics, and the order in which the quantifiers appear makes a significant difference in what the statement claims. For example, the statements "$\exists y \in \mathbb{R} \ni \forall x \in \mathbb{R}$ where $x \neq 0, xy = 1$" and "$\forall y \in \mathbb{R}, \exists x \in \mathbb{R}$ where $x \neq 0, xy = 1$" make very different claims.

The first claims that there is a real number such that if you take any real number other than zero and you multiply it to your first number, you will get

1. This is not a true statement since there is no such real number (that if you multiply it by any non-zero real number you will always get a result of 1).

The second statement claims that for all real numbers, you can find a non-zero real number whose product with your original number yields 1. This is also not true but the reason it fails is different from that of the first statement. The reason the second fails is because 0 is a real number and since the second statement makes a claim about all real numbers, the claim must hold for all real numbers and that includes 0. But, if $y = 0$, there is not a non-zero real number x for which $xy = 1$.

In general when a statement involves different types of quantifiers (universal and existence), changing the order of the quantifiers changes the statement. The new statement is hardly ever logically equivalent to the first and so it may or may not have the same truth value as the original statement. In the last example, you saw two statements which, when reversing the order of the quantifiers, yielded the same truth value, but consider also the following two statements:

$$\exists y \in \mathbb{R}^+ \ni \forall x \in \mathbb{R} \text{ where } x \neq 0, \frac{y}{x} = 1.$$
$$\forall y \in \mathbb{R}^+, \exists x \in \mathbb{R} \ni x \neq 0, \text{ and } \frac{y}{x} = 1.$$

The first of these two statements is false as there is not a positive real number y such that if you divide it by any non-zero number, you will always get a result of 1. The second statement however is true, for it is true that if you take any positive real number, you can find a non-zero number such that if you divide the two you will get a result of 1. In this case, you would simply choose the value of x to be the same as y.

So, $\exists \forall$ statements make a claim that "there is one that works for all", while $\forall \exists$ make a different claim and that is that "For each, you can find one that works."

In Class Activity 3: Determine if each of the following statements is true or false and explain your reasoning:

1. $\exists x \in \mathbb{Z} \ni \forall y \in \mathbb{N}, x < y$.
2. $\exists x \in \mathbb{Z} \ni \forall y \in \mathbb{R}, x < y$.
3. $\forall x \in \mathbb{Z}, \exists y \in \mathbb{N} \ni x < y$.
4. $\forall x \in \mathbb{R}, \exists y \in \mathbb{R} \ni x + y = 1$.

In Section 1.4 we discussed the negation, contrapositive, and converse of "if, then" statements. While the negation of a statement can be formed for any type of statement, the "contrapositive" and "converse" of a statement only have meaning in the context of statements that can be written in the form of an implication — "if p, then q." Note however that a $\forall$ statement can be written in such a form. For example, the statement "For all integers x, $x + 1$ is an integer" could also be written as "If x is an integer, then $x + 1$ is an integer." And since we recall that the negation of "If p, then q" is "p and not q", the negation of our statement above would be "x is an integer and $x + 1$ is not an integer." Note that this new statement (the negation of our original statement) claims the existence of an integer called x which has the property that $x + 1$ is not an integer. Hence, the negation of our original statement could be restated as "there is an integer x such that $x + 1$ is not an integer."

Note then that the negation of a $\forall$ statement yields a $\exists$ statement. That is, the negation of a statement of the form $\forall x, P(x)$ would be of the form $\exists x \ni \neg P(x)$. Similarly, if a statement such as "There is an $x \in \mathbb{R}$ such that x is even" is false, then that means that there is no such x, and hence "For all $x \in \mathbb{R}$, x is not even" must be true. That is, the negation of a $\exists$ statement, such as $\exists x \ni P(x)$, would be a $\forall$ statement. In the case of $\exists x \ni P(x)$, the negation would be $\forall x, \neg P(x)$.

In Class Activity 4: Negate each of the following statements:

1. $\forall x \in \mathbb{Z}, \sqrt{x} \in \mathbb{Z}$.

2. $\exists x \in \mathbb{Z} \ni x^2 \in \mathbb{Z}$.

3. $\forall x \in \mathbb{Z}, \exists y \in \mathbb{Z} \ni xy = 0$.

4. $\exists y \in \mathbb{Z} \ni \forall x \in \mathbb{Z}, xy = 0$.

5. If a divides b and a divides c, then for all integers x and y,

   a divides $bx + cy$.

Homework 1.5:

Identify the quantifiers in the following statements and determine whether each of the following statements is true or false. Explain the reasoning behind each answer.

1. $\forall x \in \mathbb{Z}, x^2 \in \mathbb{Z}$.

2. $\forall x \in \mathbb{Z}, \sqrt{x} \in \mathbb{Z}$.

3. $\exists x \in \mathbb{Z} \ni x^2 \in \mathbb{Z}$.

4. $\exists x \in \mathbb{Z} \ni \sqrt{x} \in \mathbb{Z}$.

5. $\exists! x \in \mathbb{Z} \ni x + 3 = 0$.

6. $\exists! x \in \mathbb{Z} \ni x^2 - 4 = 0$.

7. $\forall x \in \mathbb{Z}, \exists y \in \mathbb{Z} \ni xy = 0$.

8. $\exists y \in \mathbb{Z} \ni \forall x \in \mathbb{Z}, xy = 0$.

9. $\exists x \in \mathbb{Z} \ni \frac{1}{x} \in \mathbb{R}$.

10. $\exists! x \in \mathbb{R} \ni \frac{1}{x} \in \mathbb{Z}$.

11. $\exists x, y \in \mathbb{Z} \ni \frac{x}{y} = 21$. Note: $\exists x, y \in \mathbb{Z}$ means "there exists integers x and y."

12. $\exists x, y \in \mathbb{Z}^+ \ni \frac{x}{y} \in \mathbb{Z}^+$.

13. $\forall x, y \in \mathbb{Z}, \frac{x}{y} \in \mathbb{R}$.

14. $\forall x \in \mathbb{Z}^+$, if $x < 4$, then $\exists a, b, c \in \mathbb{Z} \ni a^2 + b^2 + c^2 = x$.

15. $\forall x \in \mathbb{Z}, \exists y \in \mathbb{Z} \ni x < y$.

16. $\exists x \in \mathbb{Z} \ni \forall y \in \mathbb{Z}, x < y$.

17. $\forall x \in \mathbb{R}$, if $x \neq 0$, then $\exists y \in \mathbb{R} \ni xy = 1$.

18. $\exists y \in \mathbb{R} \ni \forall x \in \mathbb{R}$ where $x \neq 0, xy = 1$.

Negate each of the following:

19. $\forall x \in A, x \geq 3$.

20. $\exists x \in \mathbb{N} \ni x \geq 3$.

21. $\forall y \in \mathbb{Z}, \exists x \in \mathbb{Q} \ni (x + 5 = y)$.

22. $\exists x \in \mathbb{R}$ such that $x^2 - 2x + 3 = 0$.

23. If x is a positive number, then there is an $\epsilon > 0$ such that $x < \epsilon$ and $\frac{1}{\epsilon} < x$.

24. Someone didn't do their homework.

25. Everything in the store is on sale.

26. $\forall x \in \mathbb{R}, \exists y \in \mathbb{R} \ni x + y = 0$.

27. $\forall x \in \mathbb{R}$, $x > 0$ implies that $\exists y \in \mathbb{R}$ such that $y < 0$ and $xy > 0$.

28. If $\exists \, A, B$, and $C \ni A \cap B = \varnothing$ and $A \cup B = C$, then $A = \varnothing$ or $B = \varnothing$.

More negations:

29. To say a function f is 1-1 means that for any elements x and y of the domain, $f(x) = f(y)$ implies $x = y$. State what it means to say that a function f is not 1-1.

30. A sequence $\{a_n\}$ is called Cauchy iff for each $\varepsilon > 0$, there is an $N \in \mathbb{N}$ such that if $m, n > N$, then $|a_m - a_n| < \varepsilon$. State clearly what it would mean for a sequence to not be Cauchy.

31. A sequence is said to be monotonically increasing iff $a_{n+1} \geq a_n$ for every index n. State clearly what it would mean for a sequence $\{a_n\}$ to not be monotonically increasing.

Chapter 2

Proof Writing

2.1 Terminology and Goals

We have already mentioned that the goal of this course is to help you develop an understanding of and an ability to construct mathematical proofs. One thing you will learn about proof writing is that it is indeed specialized writing. The object is to write out a coherent argument which could be understood by a peer — ideally someone with the same basic knowledge of mathematics as you, but who has not yet put the thought into why the statement you have proven is indeed true. Hence, you are writing the proof to explain the details to them so that they will understand. In order to write a proof, you must first believe yourself that the statement is true and have worked through the explanation in detail on your own. That is, a proof should only be written after you are certain you understand that a statement is true as well as why it is true. A proof is not an attempt to determine whether or not something is true, but rather a polished argument that is written to explain the validity of the statement to someone else — a sceptic observer.

Proofs should always be written using correct grammar and mathematical notation, and should follow a specific train of thought through to the end. Proofs should not contain side notes or details which are not needed for the argument, since things such as these detract the reader from the point of the proof.

In this course, we will look at many mathematical statements. We'll think about what they say and see if we can determine if they are true or false. Once we do so, we will try to articulate why the statement is true or false and then will write up an argument showing an outsider how we came to our decisions. A proof is a polished piece of final writing and hence should always go through the revision process. In fact, we recommend that you first write at least one draft in which you detail your thought process and then go back through that draft, reading your argument out loud as if you were explaining the result to a peer. Only then should you attempt to write up your final proof.

Now, before we actually write some proofs, a few terms and symbols are in order:

A set S is said to be **closed under an operation** $*$ iff $\forall x, y \in S, x * y \in S$.

For example, the integers are closed under addition since if you take any two integers and add them, the result will be an integer. The integers are not closed under division, since it is not the case that if you take any two integers and divide them that you will get an integer (for example $\frac{1}{3} \notin \mathbb{Z}$).

If you know that the integers are closed under addition, it is fairly straightforward to see that they are also closed under multiplication since multiplication problems can be written as addition problems (for example $5 \cdot 3 = 5 + 5 + 5$).

What about other sets under addition? Is $\mathbb{R}$ closed under addition? What about $\mathbb{Q}$ under addition? What about these sets under subtraction or under division?

We are working our way toward writing out explanations of things such as these and so continue with a few more terms.

Proof. A complete and concisely written argument establishing the truth of a mathematical statement.

Disproof. A complete and concisely written argument establishing that a statement is not true.

Theorem. A statement which has been proven to be true. The term 'theorem' indicates that it is an important result.

Conjecture. A statement which has not yet been proven, but which is believed to be true.

Lemma. A statement which is true, and which has been proven to be true, but which is not a major result in and of itself. Rather, it is a statement which is proven to be true specifically because the result will be needed to complete the proof of a subsequent theorem.

Corollary. A result which immediately follows from a theorem (usually because it is a special case of a more general result or because the proof of the corollary would directly follow from the proof of the theorem).

Definition. An established, generally accepted, meaning of a mathematical term.

Axiom. A statement which is taken to be true, without proof.

Axioms are seen as the building blocks of any system. An axiom is a statement that the mathematical community agrees to be true, but also to be so fundamental that they need not be proven. An example of an axiom is "Addition of real numbers is commutative." That is, if a and b are real numbers then $a + b = b + a$. This is a property that is so fundamental that you can't prove it. Addition is simply defined to have the commutative property and so this is considered to be an axiom. In defining any number system (such as using addition with the real numbers) you have to set up some basic rules (axioms) and definitions and then build upon those by proofs.

In this class we will take some properties as axioms and we will try to clearly state those so that you are well aware what properties must be proven, and what properties we are assuming to be true. Here are a few that we will need to use early on in the course:

Axiom 1: Addition and multiplication of real numbers is commutative. That is, $\forall x, y \in \mathbb{R}, x + y = y + x$ and $\forall x, y \in \mathbb{R}, x \cdot y = y \cdot x$.

Axiom 2: The set of real numbers is closed under addition and multiplication.

Axiom 3: The set of integers is closed under addition.

Axiom 4: For any real number x, $x + 0 = x$ and $x \cdot 1 = x$.

Axiom 5: $\forall x \in \mathbb{R}$ if $x \neq 0$, then $\frac{1}{x} \in \mathbb{R}$.

Axiom 6: Addition (and multiplication) of real numbers is associative. That is, $\forall x, y, z \in \mathbb{R}, (x + y) + z = x + (y + z)$.

Axiom 7: Multiplication is distributive over addition. That is, $\forall x, y, z \in \mathbb{R}, x(y + z) = xy + xz$.

Axiom 8: If $x = y$ and $z \in \mathbb{R}$, then $x + z = y + z$, and $xz = yz$.

Axiom 9: $\forall x, y \in \mathbb{R}, xy = 0$ iff $x = 0$ or $y = 0$.

Axiom 10: $\forall x, y \in \mathbb{R}, xy > 0$ iff $x > 0$ and $y > 0$, or $x < 0$ and $y < 0$.

Temporary Axiom: The integers are closed under multiplication.

In general, axioms are very simple properties that can be taken to be true, without proof; they are considered the building blocks of any mathematical system. You may have noted that we listed the last axiom above as a temporary axiom. This is because, once we cover the proof technique in Section 2.4 (Using Cases in Proofs), we could easily prove that the integers are closed under multiplication. However, until we get to that point, we want you to take this property as a given truth.

Regarding the axioms we have listed above, you need not try to memorize these, but rather we offer them to show you the types of things that you may assume in your proof writing. There are a number of additional things we will take for granted (and hence will take as axioms). Some of these include properties of addition, multiplication, division, and rules of exponents. For example, we will assume without proof that the way one adds two rational numbers is by getting a common denominator, and that to divide two rational numbers, one multiplies the first by the reciprocal of the second. If you have doubts about what is acceptable to assume and what should be proven, you should feel free to ask your instructor. Learning what may be taken without proof and what must be detailed is a skill you will develop over time.

Definitions are standard meanings of mathematical terms and, as such, are actually iff statements. For example, the mathematical definition of what it means for an integer to be even is as follows:

An integer x is **even** if and only if $\exists k \in \mathbb{Z} \ni x = 2k$.

However, while all definitions are indeed true iff statements, you should be aware that many mathematicians (and so many textbooks), when introducing a definition, only state them as a single implication. For example, a text might state: An integer x is said to be **even** provided that $\exists k \in \mathbb{Z} \ni x = 2k$. Now note that this statement literally only states that "If $\exists k \in \mathbb{Z} \ni x = 2k$, then is x an even integer." This is true, but the converse is also true. That is, since this is a definition, it is also true that "if x is even then $\exists k \in \mathbb{Z} \ni x = 2k$."

Technically, all definitions should be stated as iff statements but unfortunately, this is not always the practice. In this book we will make a special effort to write definitions using "iff" form, however, you should be aware that even if a definition is not written in this form it could (and should) be.

We have defined what it means for an integer to be even and now we add a few more definitions pertaining to integers:

An integer x is said to be **odd** iff $\exists k \in \mathbb{Z} \ni x = 2k + 1$.

An integer y is said to be **divisible** by an integer x iff there is an integer k such that $xk = y$.

An integer y is said to be a **multiple** of an integer x iff there is an integer k such that $xk = y$.

An integer x is said to divide an integer y ("x **divides** y") iff there is an integer k such that $xk = y$.

As an example of this last definition, 5 divides 10 since we know that $5 \cdot 2 = 10$. That is we know that there is an integer (it happens to be 2) such that 5 times that integer equals 10. The mathematical notation for "x divides y" is $x|y$. So $5|10$ and $3|33$, but $6 \nmid 21$. This last symbol "$\nmid$" represents "does not divide", just as "$\notin$" represents "is not an element of."

A word about the use of the vertical bar to write "x divides y": When you write "$x|y$" it is a shorthand notation for "x divides y", it is not indicating the operation of division. That is $x|y$ is not the same as writing "$\frac{x}{y}$" or "x/y". These later two notations indicate that you are dividing x by y and so if we return to our case of $x = 5$ and $y = 10$, it would make perfect sense to write "$10/5 = 2$" but it would be incorrect to write "$5|10 = 2$." This is because "$10/5 = 2$" states that "10 divided by 5 equals 2" which is a true statement, but "$5|10 = 2$" states that "5 divides 10 equals 2" which is not a statement. In fact, it's not even a complete sentence.

In Class Activity 1: For each of the three statements provided below, work at least five examples to help you determine whether you believe each to be true or false. If you believe the statement to be true, so state. If you believe it to be false, explain why. Also, for each statement, write the negation, the converse, and the contrapositive. Assume that x, y, and $z \in \mathbb{Z}$, and that $x \neq 0$.

1. If $x|y$ and $x|z$, then $x|(y + z)$.
2. If $x|(yz)$ then $x|y$ and $x|z$.
3. If $x|(yz)$ then $x|y$ or $x|z$.

Now we offer another important mathematical definition:

For integers a, b, and $n \in \mathbb{N}$, we say "**a is congruent to b modulo n**" iff $n|(a - b)$.

As an example of this last definition, 3 is congruent to 21 modulo 9 since $9|(3 - 21)$. We know that $9|(-18)$ since we know that $\exists k \in \mathbb{Z} \ni 9k = -18$, that integer happens to be -2. The mathematical notation for "a is congruent to b modulo n" is $a \equiv b \bmod n$ and, we are sure it is not big surprise that the notation for "a is not congruent to b modulo n" is "$a \not\equiv b \bmod n$."

In Class Activity 2: Explain why each of the following are true:

1. $11 \equiv 5 \bmod 6$
2. $3 \equiv 23 \bmod 5$
3. $0 \equiv 3 \bmod 3$
4. $5 \equiv 5 \bmod 21$
5. $5 \equiv 11 \bmod 6$

Before we proceed, we remind you of a definition that was introduced in Chapter 1. That is the definition of a rational number. Since we will use this definition quite a bit, we simply restate it here.

A real number x is said to be **rational** iff $\exists a, b \in \mathbb{Z}, b \neq 0, \ni x = \frac{a}{b}$.

Now that we have some mathematical terms defined we can begin to work on actually constructing some proofs, but first a few stylistic comments.

First off (and we can't emphasize this enough), proof writing is indeed writing and while it is appropriate to use mathematical symbols in your proofs, as stated before, the writing must be grammatically correct. That means proofs should be written using complete sentences and appropriate punctuation as well as correct mathematics. Toward this end, it is a good practice, particularly as you are first starting to write proofs, that you read your writing out loud to make sure that each and every sentence makes sense and that your ideas flow logically.

Mathematicians use special notations in order to make it easy for their readers to follow where they begin and end a proof. The two most widely used notations to let a reader know that a proof is completed are "Q.E.D." and the symbol "■". Q.E.D. represents the Latin phrase "quod erat demonstrandum" which means "which was to be demonstrated" and the shaded box gives the reader (or writer) the sense that the holes have been filled in and the argument is complete. While there are also are other variations on how to indicate the start or completion of a proof, the approach we will use in this text is to write "**Proof.**" to indicate the beginning, and "■" to indicate the end.

Another common notation that mathematicians use in proof writing is the symbol "∴". This is a shorthand notation for the word "therefore."

Additionally, you should know that if you type a proof you should italicize any and all variables so that your reader does not confuse a variable with a word. Also, if you are working through a long series of equations, think about whether or not it would help your reader if you use line breaks with your equations in such a fashion so that each line represents a new transformation.

As an example we demonstrate how the line breaks and/or use of italics helps to improve the readability of a proof. In what follows we write the exact same proof in three ways. The first is written without using italics for variables in a typed text, the second is written with an awkward line break, and the third, is written by italicizing variables and using appropriate line breaks. We trust that this will help you see how important these small details can be.

1. **Proof.** Suppose a,b, and c are integers and that a|b and a|c. Then ax=b and ay=c where x and y are integers. Then bx+cy=axx+ayy=a(xx+yy). Thus, since xx+yy is an integer, a|(bx+cy). ■

2. **Proof.** Suppose a, b, and c are integers and that $a|b$ and $a|c$. Then $ax = b$ and $ay = c$ where x and y are integers. Then $bx + cy = axx$ $+ayy = a(xx + yy)$. Thus, since $xx + yy$ is an integer, $a|(bx + cy)$. ■

3. **Proof.** Suppose a, b, and c are integers and that $a|b$ and $a|c$.
   Then $ax = b$ and $ay = c$ where x and y are integers.
   Then $bx + cy = axx + ayy = a(xx + yy)$.
   Thus, since $xx + yy$ is an integer, $a|(bx + cy)$. ■

We should also note that it is standard that proofs be written using the pronoun "we" rather than the pronoun "I". Think of it as if you are not just showing someone what you did, but rather you are trying to guide them along to make this discovery with you.

Finally, you should always begin your proof with a clear statement of all your assumptions as well as any necessary variable definitions. Also, regarding variables, in a proof, any and all variables should be defined at the time they are introduced. That is, if you introduce a variable x, you should state what you mean for x to represent. Is it an integer, a real number, a complex number, or something else?

Construction of proofs is at the very core of mathematics as this is the way mathematics advances. As with anything else we must begin with some basic building blocks (definitions, axioms, and accepted rules of logic) and from those we prove statements. Now those newly proven statements are known truths and so we can use those as we continue to prove more statements. Each question we ask and answer definitively causes us to ask more questions and so there is always new mathematics to discover.

J.J. Price published an article[1] in which he details some common mistakes that he has seen students make in their mathematics homework. The topics he covers are quite relevant to issues that arise in the writing of proofs and he provides some very good advice regarding alternatives that you might consider to avoid these mistakes. You are encouraged to read this article.

---

[1]J.J. Price, "Learning Mathematics Through Writing: Some Guidelines," *College Mathematics Journal* 20, no. 5 (1989): 393-401.

2.2   Existence Proofs and Counterexamples

We will now begin to look at several types of logical arguments that are used in proof writing, beginning with the simplest type of proof, the proof of a "there exists" statement. We do not mean to imply that all existence proofs are easy, but rather the logic behind the proof structure is rather straightforward.

Suppose that you were asked to prove that $\exists x \in \mathbb{R} \ni \frac{x}{x+1} = 5$. Since this is an existence statement, if you can demonstrate a real number with the desired property you will have proven the statement. But, remember, you must use complete sentences!

Here is a perfectly acceptable proof:

Proof. Note that $\frac{-5}{4} \in \mathbb{R}$ and $\frac{\frac{-5}{4}}{\frac{-5}{4}+1} = \frac{-5}{-5+4} = 5$. ∎

Let's try another: Prove that $\exists x \in \mathbb{Z} \ni x^2 + x + 1 > 21$.
In this case there are many such integers but for a proof you need only show that there is one. In fact you should only show that there is one, as showing that there is more than one such integer would be proving something different than what was asked, and hence would be confusing to your reader.

Here's an acceptable proof:

Proof. Note that $5 \in \mathbb{Z}$ and $5^2 + 5 + 1 = 31 > 21$. ∎

If you were reading this proof out loud you would state something along the lines of "Note that 5 is an integer and 5 squared, plus 5, plus 1, equals 31, which is greater than 21."

You may have noted that the proof doesn't include the process that we used to figure out what integer or real number had the desired property but rather it demonstrates that the number we have in mind satisfies the condition. Finding out an appropriate value is our preliminary scratch work. This brings up an important point about proof writing: You must first believe the statement is true before you try to prove it and that means that you work with the equations first.

> **In Class Activity 1:**   Prove that there exists an integer $x > 2$
> such that $x^2 - 5x + 6 = 0$.

Note that this technique of demonstrating a single example to show a statement is true only works if the quantifier is an existence quantifier. To elaborate on this, suppose we wanted to prove that $\forall x \in \mathbb{Z}, x \geq 5$, $x^2 + x + 1 > 21$ or, equivalently, "If x is an integer greater than or equal to 5, then $x^2 + x + 1 > 21$."

In this situation we must show that not only does 5 make the inequality hold true, but that every integer greater than 5 does as well. In such a case we do not want to try to prove the statement by showing how each number works individually (our proof would be infinitely long) but rather we need a general argument. We'll discuss this issue in the next section.

But, before we leave existence proofs, a note about disproofs. Since the negation of a $\forall$ statement is a $\exists$ statement, if you wanted to prove that a $\forall$ is false (that is disprove it), you could do this by making an argument that the negation of the statement is true. And indeed, to disprove a ∀ statement, one typically demonstrates an example of where the statement fails. That is, we would demonstrate that "there exists" a value for which the statement fails and hence the statement is not true "for all." That is, it is a false statement.

For example, the mathematical statement "If x is a real number, then x is an integer" is false since this means that all real numbers must be integers and we know of a real number which is not an integer ($\frac{1}{2}$ is one such number). An example which demonstrates that a statement is false is called a **counterexample**. Producing a counterexample demonstrates that a statement is not true, and hence (since it's a statement), that it is false.

A counterexample is particularly helpful if you are trying to demonstrate that a conditional statement (which can be written as a ∀ statement) is false.

> **In Class Activity 2:**   Each of the three conditional statements listed below is false. Demonstrate an example of where each of the following statements fail. That is, provide a counterexample for each statement. Note that a single, concrete example is what is desired. Once you have found a counterexample, write a disproof of the statement (hence a complete sentence showing your reader that the statement is false).
>
> 1. $\forall x \in \mathbb{Z}, \frac{x}{3} \in \mathbb{Z}$.
> 2. $\forall x \in \mathbb{Z}, \sqrt{x} \in \mathbb{Z}$.
> 3. $\forall x \in \mathbb{Z}, \frac{1}{x} \in \mathbb{R}$.

This method of presenting a counterexample works well if you want to show a conditional statement is false but it is not very helpful if you want to show that a ∃ statement is false. Why? Because for a ∃ to be false means that the condition claimed must fail for all, not just for a single value.

For example, we can not show that the statement $\exists x \in \mathbb{Z} \ni x^2 = -2$ is false by a single example; rather to explain why it is false we must make a general argument as to why all integers fail to have the property that their square is equal to -2. It is to this type of argument that we will turn our attention in the next section.

Homework 2.2:

1. Let $f(x) = 3x + 1$ and $g(x) = 6x + 5$. Prove that $\exists x \in \mathbb{R}$ such that $f(x) = g(x)$.

2. Prove that $\exists x \in \mathbb{Q}^+ \ni 4x^2 + 12x = 7$.

3. Prove that $5|20$. That is, prove that $\exists k \in \mathbb{Z} \ni 5k = 20$.

4. Prove that $56 \equiv 20 \bmod 12$.

5. Prove that $\exists x, y \in \mathbb{Z}$ such that $\frac{x}{y} = 2$.

6. In the last section you considered each of the following statements: Assume that x, y, $z \in \mathbb{Z}$, and that $x \neq 0$.

   (a) If $x|y$ and $x|z$, then $x|(y + z)$.

   (b) If $x|(yz)$ then $x|y$ and $x|z$.

   (c) If $x|(yz)$ then $x|y$ or $x|z$.

   You probably discovered at that time that the first of these statements is true while the latter two are false. Now that we have discussed disproving statements (i.e. proving a statement is false by presenting a counterexample), write a disproof for statements b) and c) above.

7. Disprove the following statement: If x and y are nonnegative integers, then $\sqrt{x + y} = \sqrt{x} + \sqrt{y}$.

8. Disprove the following statement: If x is irrational, then x^2 is irrational.

9. Provide a counterexample to show that the following statement is not true: Let $a \in \mathbb{Z}$. If $a^2 \equiv 4 \bmod 5$, then $a \equiv 2 \bmod 5$.

10. Disprove the following statement concerning integers a, b, and c: If $a|bc$, then $a|c$.

11. Disprove the following statement: If $x, y \in \mathbb{Z}$, then $3x + 2y = 1$.

12. Let $a, b, c \in \mathbb{Z}$. Disprove the following statement: If $a < b$, then $ac < bc$.

13. Let $a, b, c, d \in \mathbb{Z}$. Disprove the following statement: If $a > b$ and $c > d > 0$, then $ac > bd$.

2.3 Direct Proofs ("If, then" or "For every" Statements)

Implications or "∀" statements make conditional claims. If the hypothesis is met, then the conclusion must follow. In order to prove such a statement in a straightforward manner, one must make a general argument explaining why if the hypothesis is met, the conclusion must occur.

We begin by some examples. Recall that an integer x is said to be even iff $\exists k \in \mathbb{Z} \ni x = 2k$, and also recall that we know that the integers are closed under multiplication.

Consider the following statements along with a valid proof of each:

Statement. If x is an even integer, then x^2 is even.

Proof. Let x be an even integer.
Then $\exists k \in \mathbb{Z} \ni x = 2k$.
So we know $x^2 = (2k)^2 = 4k^2 = 2(2k^2)$.
Now since the integers are closed under multiplication,
we know that $2kk$ is an integer.
That is, $q = 2k^2$ is an integer such that $x^2 = 2q$.
Therefore, x^2 is even. ∎

Statement. The rational numbers are closed under addition.

Proof. Let x and y be rational numbers.
We know that $\exists a, b, c, d \in \mathbb{Z}$, with $b \neq 0$ and $d \neq 0 \ni x = \frac{a}{b}$ and $y = \frac{c}{d}$.
Note that $x + y = \frac{a}{b} + \frac{c}{d} = \frac{ad+bc}{bd}$.
And since we know that the integers are closed under multiplication,
we know that ad, bc and bd are all integers.
Furthermore, we know that since the integers are closed under addition,
$ad + bc$ must be an integer.
Hence we have shown that $x+y$ can be written as the quotient of two integers
∴ $x + y$ is a rational number.
That is, the rationals are closed under addition. ∎

Note that in both of these examples, the proof began by defining variables which met the criteria of the hypothesis. Recall that an "if, then" statement will be true if you know that when the hypothesis is true the conclusion is also true. Hence, direct proofs of "if, then" statements should begin with a statement of what is given in the hypothesis.

So, in such a proof, we begin with the hypothesis, and use what we already know to show that the conclusion must follow. Sometimes this is an easy process, but most often it is not. By this we mean that very often it will be the case that you do not easily see how the hypothesis leads to the conclusion. Hence, often in proof writing we begin by working out those details. This, however, is not a proof — a proof is a polished piece of work that shows the connection. It should give your reader no idea as to how easy or hard this was for you to work through and it should not include any wrong paths or other ideas that you had which are not necessary for your argument.

Consider the following statement: If x and y are even integers, then $x + 3y$ is an even integer.

In designing a proof of this statement we would begin by supposing that the condition of the hypothesis is met. That is, "suppose x and y are even integers." Then we would want to conclude from this (and any other information we had at our disposal) that $x + 3y$ must be even. So, thus far we know that our proof will look something like the following (using an open box, $\square$, to indicate that though we have written the last line of the proof, we recognize that our proof is not complete):

Proof. Suppose x and y are even integers.

   ...

   Therefore $x + 3y$ is even. $\square$

Now in order to connect the conclusion to the hypothesis we know that we are going to have to show that there is an integer k such that $x + 3y = 2k$ since this is what it means for $x + 3y$ to be even. Also, we know that since x and y are both integers then there exists integers a and b such that $x = 2a$ and $y = 2b$. Note that we choose two different variables (a and b) since if we used the same variable for both x and y then this would require that x and y be equal which is not part of the hypothesis. So we now can extend our proof draft to the following:

Proof. Suppose x and y are even integers.
   Then $\exists a, b \in \mathbb{Z} \ni x = 2a$ and $y = 2b$.
   ....

   Thus $\exists$ an integer k such that $x + 3y = 2k$.
   Therefore $x + 3y$ is even. $\square$

As we seek to see how we can end up with an equation of the form $x + 3y = 2k$, the easiest thing to do would be to start by writing $x + 3y$ and then rewriting this in a form which contains a factor of 2, that is $x + 3y = 2a + 3(2b)$. Note now that this latter form could be rewritten (using only our axioms relating to commutativity and distribution of real numbers) as $2(a + 3b)$. And, as the integers are closed under multiplication and also under addition, we know that the quantity $a + 3b$ must be an integer. Thus we have found an integer, call it k, such that $x + 3y = 2k$.

Now that we have gone through the entire thought process, we offer an actual final proof:

Proof. Suppose x and y are even integers.
   Then $\exists a, b \in \mathbb{Z} \ni x = 2a$ and $y = 2b$.
   Consider $x + 3y$.
   Substituting for x and y yields, $x + 3y = 2a + 3(2b) = 2(a + 3b)$.
   Now since we know that the integers are closed under multiplication,
   $3b$ must be an integer.
   Also, since we know that a and $3b$ are both integers
   and since we know that the integers are closed under addition,
   we know that $a + 3b$ must be an integer.

Hence, $\exists$ an integer k such that $x + 3y = 2k$ (the specific value of k is $a + 3b$). Therefore $x + 3y$ is even. ∎

Consider the following statement along with an attempted proof of that statement.

Statement. For any positive real number x, $x + \frac{1}{x} \geq 2$.

Proposed Proof.   Let x be a positive real number.

Multiplying both sides of the inequality $x + \frac{1}{x} \geq 2$ by x
yields $x^2 + 1 \geq 2x$, which we can rewrite as $(x-1)^2 \geq 0$.
Since this equivalent expression is true,
the original inequality must be true. ∎

While the statement provided above is indeed true, the proposed proof is incorrect. To see this, note that the statement can be rewritten as the implication: If x is a positive real number, then $x + \frac{1}{x} \geq 2$.   In the proposed proof we see that the author began by assuming the hypothesis, which means they are indeed starting off with a valid, direct, line of reasoning. However, in their second sentence when they tell their reader to multiply both sides of the inequality by x, we see that they are assuming that $x + \frac{1}{x} \geq 2$. is valid. That is, in their proof, they are assuming the conclusion of the statement to be true. This renders the proof invalid.

Be careful whenever you are writing a direct proof that you do not assume the conclusion to be true, but rather build up an explanation of why it must be true.

In Class Activity:

1. Prove that the sum of two even numbers is even.

2. Prove that the product of two odd numbers is odd.

3. Prove that if $\sqrt{x} \in \mathbb{Q}$, then $x \in \mathbb{Q}$.

4. Let a, b, and c, be non-zero integers. Prove that $ac|bc$ iff $a|b$.

Homework 2.3:

1. What would you assume (and hence what should be the first portion of your proof) if you were proving the statement "If $x + y$ is even and y is odd, then x is odd" via a direct method?

2. What would you assume if you were proving the statement "If $x^2 + y = 13$ and $y \neq 4$, then $x \neq 3$" via a direct method?

3. Consider the following statement together with a proposed proof of that statement. If the proof is correct (in that it proves the statement given), so state. If the proof is incorrect, state one significant problem that you have identified.

   Statement. Let $a \in \mathbb{Z}$. If 2 divides a and 3 divides a, then 6 divides a.
   Proposed Proof. Let $a \in \mathbb{Z}$ and assume that
       2 divides a and 3 divides a.
       Since 3 divides a, we know that $3k = a$ for some integer k.
       Note that if k is even then $k = 2q$ for some integer q
       and so $a = 3k = 3(2q) = 6q$. Thus 6 divides a. ∎

4. In the last section we brought back to your attention three statements:
   Assume that x, y, and $z \in \mathbb{Z}$, and that $x \neq 0$.
   i) If $x|y$ and $x|z$, then $x|(y + z)$.
   ii) If $x|(yz)$, then $x|y$ and $x|z$.
   iii) If $x|(yz)$, then $x|y$ or $x|z$.

   You disproved statements ii and iii in the last section and, as you know, statement i is true. We now ask that you write a proof of statement i.

5. Prove that if a is an integer then $2a - 3$ is odd.

6. Prove the following: Let $x, y \in \mathbb{Z}$. If $x + y$ is even and y is odd, then x is odd.

7. Prove the following: Let $a, b, c \in \mathbb{Z}$ with $a \neq 0$ and $b \neq 0$. If $a|b$ and $b|c$, then $a|c$.

8. Prove the following: Let $a, b \in \mathbb{Z}$. If $a|b$, then $a|(5b - 2a)$.

9. Consider the following statement along with an incorrect proposed proof of that statement:

   Statement. If n^2 is even, then n is even.
   Proposed Proof. Suppose n^2 is even. Then $n^2 = (2k)^2 = (2k)(2k)$.
       Also we know $n^2 = (n)(n)$. Therefore $n = 2k$ and so n is even. ∎

   What is wrong with this proof?

10. Let $a, b \in \mathbb{Z}$ and $n \in \mathbb{Z}^+$. Prove that if $a \equiv b \bmod n$ then $b \equiv a \bmod n$.

11. Prove that if $x \in \mathbb{Q}$, then $x - 5x^2 \in \mathbb{Q}$.

12. You have nine friends who are taking Introduction to Advanced Math, a course you took last semester. For homework in this class they were asked to prove or disprove the following statement: For $a, b, c \in \mathbb{Z}$, if $a|b$ and $a|c$, then $\forall x, y \in \mathbb{Z}$, $a|(bx + cy)$. Since you did so well in the class each friend has asked you to look over their homework and give them some feedback. For each, determine if their proof is an acceptable argument. If it is not, identify the major problem. If the proof is logically acceptable but still unclear or not detailed enough, write out some suggestions for the author.

(a) **Proposed Disproof.** Let $a = 12$, $b = 4$, and $c = 3$.
   We know that $12|4$, and $12|3$, and when $x = 1$, $y = 5$,
   $bx + cy = 4 \cdot 1 + 3 \cdot 5 = 19$.
   Therefore there exists integers x and y such
   that a does not divide $bx + cy$.
   Therefore, this statement is not true. ∎

(b) **Proposed Proof.** Assume $a|b$ and $a|c$ for integers a, b, and c.
   So $a \cdot m = b$ and $a \cdot n = c$, for some n, m .
   Then $bx + cy = amx + any = a(mx + ny)$.
   Therefore, $a|(bx + cy)$. ∎

(c) **Proposed Proof.** If $a|b$, then ab, and if $a|c$, then ca.
   Therefore, a is a multiple of both b and c,
   so a is also a multiple of $bx + cy$. ∎

(d) **Proposed Proof.** For integers a, b, and c, suppose $a|b$.
   Then $ak = b$ and $a|c$ such that $aq = c$ where k and q are integers.
   Now $b + c = ak + aq = a(k + q)$, thus $a|(b + c)$.
   Then consider $bx + cy$ for $x, y \in \mathbb{Z}$.
   $bx + cy = akx + aqy = a(kx + qy)$.
   Therefore by direct proof, $a|(bx + cy)$ if $a|b$ and $a|c$. ∎

(e) **Proposed Proof.** Suppose that $a|b$ and $a|c$.
   Note then that $a = bx$ for some $x \in \mathbb{Z}$.
   Also note that $a = cy$ for some $y \in \mathbb{Z}$.
   So $a + a = bx + cy$ or $2a = bx + cy$.
   Therefore, $a|(bx + cy)$. ∎

(f) **Proposed Proof.** Assume that $akx = bx$ and $amy = cy$.
   Since kx and my are both integers we can rewrite this statement
   as $aq = bx$ and $al = cy$.
   Therefore a divides bx and a divides cy.
   Thus $a|(bx + cy)$ for all $x, y \in \mathbb{Z}$. ∎

(g) **Proposed Proof.** Suppose $\frac{b}{a} = k$, and $\frac{c}{a} = q$.
   Then $bx + cy = akz + aqy = a(kx + qy)$.
   Therefore, $a|(bx + cy)$. ∎

(h) **Proposed Proof.** Suppose $a|b$ and $a|c$.

Then $ax = b$ and $ay = c$ for any integers x and y.

Then $bx + cy = (ax)x + (ay)y = a(xx + yy)$.

Since x and y are any integers and the integers are closed under addition and multiplication, we know that $(xx + yy) \in \mathbb{Z}$.

Thus $a|(bx + cy)$. ∎

(i) **Proposed Proof.** Suppose that $a|b$ and $a|c$.

Then $a = bk$ and $a = qc$ for some integers k and q.

Thus $bx + cy = \frac{a}{k}x + \frac{a}{q}y = a\left(\frac{x}{k} + \frac{y}{q}\right)$.

Therefore $a|(bx + cy)$. ∎

These last two statements involve double quantifiers and hence, involve both the concepts of proving a "for every" statement as well as a "there exists" statement.

13. Prove that if $y \in \mathbb{Z}$, $y \neq 0$, then $\exists x \in \mathbb{Z}$ such that $\frac{x}{y} = 2$.

14. Let $f(r) = \frac{2}{r-3}$ where $r \in \mathbb{R}$. Prove that $\forall y \in \mathbb{R}, y \neq 0$, $\exists x \in \mathbb{R}$ such that $f(x) = y$.

2.4   Using Cases in Proofs

Sometimes it is easier to prove a statement by breaking the hypothesis down into cases and arguing that the conclusion must follow in each and every case. One such example of when you might want to do this would be if you were wanting to construct a direct proof for a statement which contained an "or" condition in the hypothesis — something like "If a is divisible by 3 or if b is divisible by 3 then $5ab$ is divisible by 15." As another example, if you wanted to prove something about a real number x, it might make sense to break your proof into three cases — whether x is positive, negative, or 0. Or, if x is an integer, perhaps you might want to break your proof into cases where x is even and odd... or could you? Is it true that any integer must be either even or odd? This is something you've probably assumed for most of your life but we now explain why your intuition is accurate on this issue.

The Division Algorithm is an important result pertaining to the set of integers which is often invoked in a proof by cases pertaining to integers. The Division Algorithm can (and will be) proven once we complete the material in this chapter, but for now we simply present the result.

The division algorithm asserts that given any integers a and b with $b > 0$, there exists unique integers q and r with $0 \le r < b$, such that $a = bq + r$. Using mathematical notation this can be written as $\forall a, b \in \mathbb{Z}$ with $b > 0, \exists! q, r \in \mathbb{Z}$, with $0 \le r < b \ni a = bq + r$.

For example, suppose $a = 3$ and $b = 5$. Then the division algorithm asserts that there exists unique integers q and r with such that $3 = 5q + r$. In this case it is easy to see that $q = 0$ and $r = 3$ are integers which will satisfy this equation and they are the only integers that will satisfy it given the restriction that $0 \le r < 5$.

What if $a = 10$ and $b = 5$? In this case the unique q and r would be $q = 2$ and $r = 0$. What about $a = -21$ and $b = 5$? Here the unique q and r are $q = -5$ and $r = 4$.

> **In Class Activity 1:**   For each of the following determine the integers q and r with $0 \le r < b$ such that $a = bq + r$.
>
> 1. $a = 7$ and $b = 2$
> 2. $a = -5$ and $b = 3$

The particular relevancy of the division algorithm to us is that it allows us to break down a problem regarding integers into a distinct number of cases.

For example if you wanted to prove something about an integer x and, for reasons that will soon become clear, you wanted to consider this integer x as being written in terms of a multiple of 5, then the division algorithm allows you to do just this. From the division algorithm you can definitively know that there is an integer q such that your integer x can be written in exactly one of the following forms: $x = 5q$, $x = 5q + 1$, $x = 5q + 2$, $x = 5q + 3$, or $x = 5q + 4$.

Similarly, if you wanted to prove something about an integer x and you wanted to consider this integer x as being written in terms of a multiple of 2, the division algorithm assures you that either $\exists q \in \mathbb{Z} \ni x = 2q$, or $\exists q \in \mathbb{Z} \ni x = 2q + 1$. So in proving something about the integers you might opt to break your argument into two cases, one where x is even (of the form $2q$ for some integer q), and the other when x is odd (of the form $2q + 1$ for some integer q).

Consider the following statement along with a valid proof:

Statement.   If x is any integer, then $x^2 + x$ is even.

Proof. Let x be an integer.

We know from the division algorithm that there is an integer q such that either $x = 2q$ or $x = 2q + 1$.

If $x = 2q$, then $x^2 + x = (2q)^2 + (2q) = 4q^2 + 2q = 2\left(2q^2 + q\right)$.

Since we know the integers are closed under multiplication, $2q^2$ must be an integer, and since the integers are closed under addition, $2q^2 + q \in \mathbb{Z}$. Therefore $x^2 + x$ is even.

If $x = 2q+1$, then $x^2 + x = (2q + 1)^2 + (2q + 1) = \left(4q^2 + 4q + 1\right) + (2q + 1) = 4q^2 + 6q + 2 = 2\left(2q^2 + 3q + 1\right)$. Since the integers are closed under multiplication, $2q^2$ and $3q$ are both integers. And, since the integers are closed under addition, $2q^2 + 3q + 1 \in \mathbb{Z}$.

Therefore, $x^2 + x$ is even. ∎

In Class Activity 2:   Let $a \in \mathbb{Z}$. Prove that if $3 \nmid a$, then $3 \nmid a^2$.

In Class Activity 3:   In Section 2.1 we stated as a theorem that the integers are closed under multiplication. This fact can be proven from the axiom that $\mathbb{Z}$ is closed under addition. Use the axiom that $\mathbb{Z}$ is closed under addition to prove that $\mathbb{Z}$ is closed under multiplication. That is, prove that if $x, y \in \mathbb{Z}$, then $xy \in \mathbb{Z}$.

In Class Activity 4:   Let $a, b \in \mathbb{R}$. Prove that $|ab| = |a|\,|b|$.

Homework 2.4:

1. The statement "If x is any real number, then $x^2 > x$" is a false statement and so the following attempt at a proof by cases must be incorrect. Read the proposed proof and identify the problem with it. Once you have identified the fatal flaw of this proof, disprove the statement.

   Proposed Proof.[2] Let x be a real number.
   Case 1: If $x \leq 0$ then $x^2 \geq 0$ while $x \leq 0$, therefore $x^2 \geq x$.
   Case 2: If $x \geq 1$ then we can multiply both sides of the equation $x \geq 1$ by x to see that $x^2 \geq x$.
   Since both cases lead to the desired conclusion,
   $x^2 \geq x$ for any real number x. ∎

2. Critique the following proof. If the proof is correct (in that it proves the statement given), so state. If the proof is incorrect, state one <u>significant</u> problem that you have identified.

   Statement. Let x, $y \in \mathbb{R}$. If $xy = 0$, then $x = 0$ and $y = 0$.
   Proposed Proof. We know that either $x = 0$ or $y = 0$.
   Case 1: If $x = 0$ then $xy = 0y = 0$.
   Case 2: If $y = 0$ then $xy = x0 = 0$.
   So in all cases we see that $xy = 0$. ∎

3. Suppose that you wanted to prove that if a is an integer, then $a(a^2 + 2)$ is divisible by 3. If you wanted to set up a direct proof using cases for the possible value of a (in the form of 3 times an integer plus an integer), what would the cases be? List them all (but don't try to prove anything).

4. Consider the following statement: Let $a, b \in \mathbb{Z}$. If $5|ab$ then $5|a$ or $5|b$. One way to prove this statement is to begin by supposing that a and b are integers and that 5 doesn't divide a and also that 5 doesn't divide b (we'll talk about why this is an appropriate assumption in the next section). From this, you can proceed by looking at a number of cases that arise from your supposition. What would the cases look like and how many cases would there be? Don't prove any of them but do explain what the cases would look like and how many cases would need to be covered. Feel free to list the cases out, but you don't have to list them all, just provide a general explanation of what the cases must cover along with the number of cases that an exhaustive list should include.

5. Let $x, y \in \mathbb{Z}$. Prove that $x^2 + y^2 + x + y$ is even.

---

[2]Robert Wolf, *Proof, Logic, and Conjecture: The Mathematician's Toolbox* (New York: W.H. Freeman, 1998), 91.

6. Prove that any integer can be written in one of the following forms: $3k-1$, $3k$, or $3k+1$, where $k \in \mathbb{Z}$.

7. Prove that $\forall x \in \mathbb{R}, |x-1| = |1-x|$.

8. Prove that if n is an odd integer then $8|(n^2-1)$. *Note:* There are many right ways to do this but try it by first explaining why any odd integer n must be of the form $4k+1$ or $4k+3$ for some integer k, and then proving that in both of these cases, $8|(n^2-1)$.

9. Prove that the square of any odd integer is of the form $8m+1$ for some integer m.

2.5   Contrapositive Arguments

Recall from our earlier work that the contrapositive of an implication statement is logically equivalent to the statement. That is, "If p, then q" has the same truth value as the statement "If not q, then not p" and hence, one way to prove that an implication statement is true (or false) is to prove that it's contrapositive is true (or false).

As an example, suppose that you know that x is a real number, and you wanted to prove the following statement: If x is not an integer, then $\frac{x}{2}$ is not an integer.

You might try to prove this by a direct argument and in this case your initial assumption would be that "suppose that x is not an integer." Unfortunately this does not give you much to work with since at this point we have not proven anything about real numbers which are not integers. However, if you try to prove the statement by a contrapositive argument, then your supposition would be the negation of "$\frac{x}{2}$ is not an integer" which is "$\frac{x}{2}$ is an integer" and indeed this provides us with information which we can use. Using this, we offer the following proof of our statement:

Proof. Suppose that $\frac{x}{2} \in \mathbb{Z}$.

Then since the integers are closed under multiplication,

we know that $2 \cdot \frac{x}{2} \in \mathbb{Z}$.

That is, $x \in \mathbb{Z}$. ∎

In the proof above, we have shown "If $\frac{x}{2}$ is an integer, then $x \in \mathbb{Z}$" must be true. Furthermore, as this is logically equivalent to the statement "If x is not an integer, then $\frac{x}{2}$ is not an integer," we know that this later statement must be true as well.

Here's another classic example of a proof for which a contrapositive argument proves (pun intended) to be quite helpful. Consider the following statement: Let x be an integer. If x^2 is even, then x is even.

Note that if we tried to prove this using a direct argument our beginning attempt would probably fall along the lines of the following:

Proof. Let x be an integer and suppose that x^2 is even.

Then $\exists k \in \mathbb{Z} \ni x^2 = 2k$. ...

But at this point in the proof, we would most probably be stuck as there is no clear way to get from $x^2 = 2k$ to $x = 2q$ where we know that q is an integer. Our only hope here seems to be to try to take the square root of both sides of our equation, but this becomes $\sqrt{x^2} = \sqrt{2k}$ and so at best we have that $|x| = 2\sqrt{\frac{k}{2}}$, and we certainly have no hope of knowing that $\sqrt{\frac{k}{2}} \in \mathbb{Z}$.

So, we appear to be stuck in our attempt at using a direct argument. However, there are alternatives to using a direct argument and this statement provides us with a classic example of when the use of a contrapositive argument is helpful. Now that we (hopefully) have convinced you why the direct argument is not worthwhile, we present a valid proof of the statement using a contrapositive argument:

Proof. Let x be an integer and suppose that x is not even.

We know from the division algorithm that $x = 2q + 1$ for some integer q. Thus $x^2 = (2q + 1)^2 = 4q^2 + 4q + 1 = 2(2q^2 + 2q) + 1$ and since we know that $2q^2 + 2q \in \mathbb{Z}$, we know that x^2 is not even. ∎

Note that we did not comment in the proof on what type of argument we were making (a contrapositive argument). Mathematicians know that the contrapositive argument is another way of proving a statement and hence you do not need to add in a closing sentence in your proof which states "and so as this statement is true, it's contrapositive must also be true," although you are welcome to do so. You should however, as mentioned earlier, clearly state your beginning assumption in the first sentence or two of your proof. That is, your first few sentences should make it clear to your reader what type of argument you are using (be it direct, contrapositive, or something else).

In Class Activity:

1. Let $x, y \in \mathbb{R}$. Prove that if $x + y$ is irrational,

   then x is irrational or y is irrational.

2. Let $a, b, d \in \mathbb{Z}, d \neq 0$. Prove that if $d \nmid ab$,

   then $d \nmid a$ or $d \nmid b$.

We are now ready to focus on another type of argument, but first we offer a few exercises so that you can practice working with a contrapositive argument.

Homework 2.5:

1. What would you assume if you were proving the statement, "If $x^2 + y = 13$ and $y \neq 4$, then $x \neq 3$" via a contrapositive argument?

2. What would you assume if you were proving the statement, "If $x + y$ is even and y is odd, then x is odd" via a contrapositive argument?

3. If you proceeded to go through a complete proof of the statement in question 2 above via a contrapositive argument, what should the conclusion of your argument be?

4. Let $f(x) = 3x - 2$ for $x \in \mathbb{R}$. Prove that if $x \neq y$, then $f(x) \neq f(y)$.

5. Let $m, n \in \mathbb{Z}$. Prove that if mn is even, then m is even or n is even.

6. Let x be an integer. Prove that if x^2 is not divisible by 4, then x is odd.

7. Let $x \in \mathbb{R}$. Prove that if x is irrational, then $\sqrt[3]{x}$ is irrational.

2.6   Contradiction Arguments

A **contradiction** is a statement which can never be true. Note that this means that a contradiction is a false statement, but it is much more than that. To clarify this note that the statement "If x is an integer, then $\frac{x}{2}$ is an integer" or equivalently "Let x be an integer, then $\frac{x}{2}$ is an integer" is a false statement. But, even though this statement is false, there are *some* integer values of x for which it would be true. That is, it's a false statement even though some of the elements which satisfy the hypothesis do indeed satisfy the conclusion. Such a statement is not considered to be a contradiction. An example of a contradiction would be something along the lines of "Let x be an integer, then x is positive and x is negative." This statement is not only false, it is not true for any value which would satisfy the hypothesis.

As another example, consider the statement "x is an even integer and x is not divisible by 4." This statement is not a contradiction as it is not impossible to find a value of x for which the statement holds. That is, $x = 2$ is an example of an even integer which is not divisible by 4. However, "x is an even integer and x is an odd integer" is a contradiction as these two conditions cannot possibly occur simultaneously (a direct consequence of the division algorithm). In terms of a truth table, a contradiction would generate all "F"s. There is no circumstance under which it could be true.

Returning to our original example, "If x is an integer, then $\frac{x}{2}$ is an integer" note that any specific value of x which make the original statement true will make its negation false (and vice versa). For example, as $x = 3$ provides us with an example of where the original statement is false, $x = 3$ will necessarily provide us with an example of where the negation of the original statement is true. And as $x = 6$ provides us with an example of where the original statement is true, $x = 6$ will necessarily provide us with an example of where the negation of the original statement is false.

A proof by contradiction is a proof in which you assume that the negation of your statement is true and then show that this is a contradiction. If you show that the negation of the statement cannot under any circumstances be true, then you have in effect shown that the original statement cannot, under any circumstances, be false. That is, you will have shown that the original statement is true.

At the risk of sounding redundant, we caution that it would be insufficient to simply show an example of when the negation of the statement would be false as this would simply be an example of when the original statement is true and so would not prove that the original statement is true in general, which is what it means to say a statement is true. Recall that our false statement, "If x is an integer, then $\frac{x}{2} \in \mathbb{Z}$" is true for some values of x, but it is not always true, and hence, it is a false statement.

Now that we have explained the thought process of a proof by contradiction, we offer an example. Consider the following statement together with a valid proof:

Statement. Let $a, b \in \mathbb{Z}$. If $a \cdot b$ is odd, then a is odd or b is odd.
Proof. Suppose the statement is not true.
  That is, suppose that $a, b \in \mathbb{Z}$, $a \cdot b$ is odd, a is even, and b is even.
  Then $\exists k, q \in \mathbb{Z} \ni a = 2k$ and $b = 2q$.
  Note then that $a \cdot b = (2k)(2q) = 4kq = 2(2kq)$ and as we know $2kq \in \mathbb{Z}$,
  we know that $a \cdot b$ is even.
  But this is not possible since $a \cdot b$ is odd.
  Thus we have a contradiction and so the original statement must be true. ∎

In Class Activity:

1. Let $x, y \in \mathbb{Z}$. Prove that if x is even and y is odd,
   then $4 \nmid (2x^2 + y^2)$.

2. Prove that if $x \in \mathbb{Q}$, $x \neq 0$, and y is irrational,
   then xy is irrational.

3. Let $a, b, c \in \mathbb{R}$. Prove that if $a < b$ and $c > 0$, then $ac < bc$.

We end this section with two classic number theory theorems which are typically proven via a contradiction argument. The first is that there are infinitely many primes, and the second, that the square root of 2 (or in general, the square root of any prime number) is irrational.

A **prime number** is a positive integer greater than 1 whose only positive divisors are 1 and the number itself. We know of many prime numbers, e.g. $2, 3, 5, 7, 11, 13, 17, 19, 23, 31, \ldots$ At the time this text was written, the largest known prime number consisted of $9,808,358$ digits. Sometimes, it looks as if the gaps between primes get larger and larger and mathematicians are constantly seeking to find the next largest prime number. Will someone ever find the last one? Will there ever be a time when all the prime numbers are all known? The answer is unequivocally NO! Even though we don't know what the next largest prime number is, we know one will always exist. Here's why:

Theorem. There are infinitely many primes.
Proof. Suppose the theorem is not true. Then there are finitely many primes.
  Hence we can list them all. Suppose that $p_1, p_2, \ldots p_k$ is a complete list
  of all the prime numbers. Now note that since all these primes must be
  positive integers, $p_1 \cdot p_2 \cdots p_k + 1$ is a positive integer,
  and it is larger than any of the prime numbers. Hence it is not a prime
  number and so it can be factored into a product of primes.
  That is, there must be some prime number p_i in our list of primes
  for which $p_i | (p_1 \cdot p_2 \cdots p_k + 1)$.
  In other words, we know that $\exists q \in \mathbb{Z} \ni p_i q = p_1 \cdot p_2 \cdots p_k + 1$.
  But note that since p_i is one of the prime numbers in the listing, we can
  rewrite this last equation to yield $p_i (q - p_1 p_2 \cdots p_{i-1} p_{i+1} \cdots p_k) = 1$ which
  means that $p_i | 1$.
  However this is impossible since the only integers which divide 1
  are 1 and -1, neither of which are prime.
  Hence, there must be infinitely many primes. ∎

You have probably been told that $\sqrt{2}$ (or the square root of any prime number) is an irrational number, as is π, and as is e. But do you know why? It is a rather classic number theory proof to show that the square root of a prime number is irrational and you now have to skills to fully understand this proof. We should also note that it requires a bit more to show that π and e are irrational, but if you're curious you should definitely plan to take Abstract Algebra.

Since the definition of a real number being irrational is that it is not rational, the natural idea to prove that a number is irrational would be to try to make a contradiction argument in which you assume that the number is rational and reach a contradiction. This contradiction argument works very well to prove that the square root of a prime number is irrational. It does, however, use one theorem which we have not proven, but which can be proven relatively easily. In fact it would be among one of the first things you do in a class in number theory. However, in the interest of time we simply present it here without proof:

Theorem.   Let $a, b, p \in \mathbb{Z}$. If p is prime and $p|ab$, then $p|a$ or $p|b$.

We proved a special case of this theorem earlier in this section when we proved that if ab is even then a is even or b is even. Although we are not going to formally prove the general result here we would like to make sure that the idea makes sense. First off, it should be clear to you that the restriction that p be a prime number is necessary since this awareness really helps you to see the idea behind the proof. Note that $6|(3 \cdot 4)$ but $6 \nmid 3$ and $6 \nmid 4$. This is because some, but not all, of the factors of 6 are in 3 and some, but not all, of the factors of 6 are in 4. However, when the divisor is a prime number it is not possible to split up the factors this way. If we say $5|(a \cdot b)$, there is no way for the factors of 5 to be split up between a and b since 5 is prime. Hence either a or b must be divisible by 5.

So, without further ado we present you with one of the classic contradiction proofs:

Theorem. $\sqrt{3}$ is irrational.

Proof. Suppose that $\sqrt{3}$ is not irrational.

This means $\sqrt{3}$ is rational and hence, $\exists a, b \in \mathbb{Z}$ with $b \neq 0$ such that $\sqrt{3} = \frac{a}{b}$. Furthermore, of all the ways to write $\sqrt{3}$ as an integer divided by an integer, we can insist that this a and b be chosen so that they have no common factors greater than 1.

That is, this fractional representation of $\sqrt{3}$ is in reduced form.

Now note that squaring both sides of our equation above yields $3 = \frac{a^2}{b^2}$, and so $3b^2 = a^2$. Thus $3 | a^2$.

But since 3 is prime, we know from our previous theorem that $3 | a$.

Thus $a = 3k$ for some integer k and so our equation could be written as $3b^2 = (3k)^2$ or $b^2 = 3k^2$.

Thus $3 | b^2$. Hence $3 | b$.

But then we have that $3 | a$ and $3 | b$ and we specifically chose a and b so that they had no common factors greater than 1.

Hence we have reached a contradiction.

Thus $\sqrt{3}$ cannot be rational and hence must be irrational. ∎

Homework 2.6:

1. What would you assume if you were proving the statement "If $x + y$ is even and y is odd, then x is odd" via a contradiction argument?

2. What would you assume if you were proving the statement "If $x^2 + y = 13$ and $y \neq 4$, then $x \neq 3$" via a contradiction argument?

3. Prove that if x and y are integers, then $8x + 2y \neq 3$.

4. Let $a, b, c \in \mathbb{R}$. Prove that if $a > b$ and $c > 0$, then $ac > bc$.

5. Prove that if x is a positive real number, then $x + \frac{1}{x} \geq 2$.

6. Prove that if x is rational and y is irrational, then $x + y$ is irrational.

7. Prove that if $x, y \in \mathbb{R}^+$ such that $\sqrt{xy} \neq \frac{x+y}{2}$, then $x \neq y$.

8. Let $x \in \mathbb{R}^+$. Prove that $\frac{x}{x+1} < \frac{x+1}{x+2}$.

9. Prove that $\sqrt{2}$ is irrational.

10. Let $a, b \in \mathbb{Z}$. Prove that if $3 | (a \cdot b)$, then $3 | a$ or $3 | b$.

11. Consider the following attempt to prove the statement "If $a \in \mathbb{R}^+$, then $a > \sqrt{a}$." This proposed proof is a valid argument until you get to the last sentence. What is the issue? Can it be fixed? Why or why not?

    Proposed Proof. Suppose not.
      That is, suppose that $a \in \mathbb{R}^+$, and that $a \leq \sqrt{a}$.
      Then since a is positive, we know that we can multiply both sides of this inequality by a to get $a^2 \leq a\sqrt{a}$.
      Also since a is positive, we could multiply both sides our first inequality by $\sqrt{a}$ to get $a\sqrt{a} \leq \sqrt{a}\sqrt{a}$.
      Then, putting these together we see that
      $a^2 \leq a\sqrt{a} \leq \sqrt{a}\sqrt{a} = a$.
      Thus, $a^2 \leq a$ and so $a^2 - a \leq 0$. And so $a(a - 1) \leq 0$.
      Now since we know $a > 0$, it must be the case that $a - 1 \leq 0$.
      That is $a \leq 1$. But this is impossible and so the original statement must be true. ∎

2.7   Putting it All Together

At this point you have studied several type of arguments commonly used in proof writing. In this section there are no new techniques, but we provide a way for you to practice determining what type of argument might work best for a given statement. That is, we want you to have the opportunity to practice not only writing proofs, but also in determining what type of argument you may want to try in writing a proof of a given statement. We would like to make it clear that there are certainly statements for which more than one type of proof argument will work. That is, statements which can be proven directly can also sometimes be proven using a contrapositive argument, as well as by other means. It is not that one of these proof strategies is better than another, rather that each technique offers us a unique way to use the information in the statement we are trying to prove. Additionally, it might be that within a direct proof that you need to show some smaller detail and you do that (sub-argument) by contradiction. So we consider each of the methods we have shown to be tools you should know how to use. Pending the problem, sometimes one tool works better than another and sometimes you find that you need more than one tool to get the job done. Also, sometimes even though you can get a job done with several tools, one particular tool seems to work best. It was with this last analogy in mind that Paul Erdős used to say that he believed that God had written a document called *The Book*. *The Book*, Prof. Erdős believed, contains all the great theorems of mathematics proven in their most elegant form. Prof. Erdős died in 1996 and we hope that he is still finding great pleasure in his reading.

Before we begin our work, we recap on the ideas of Sections 1-6. We have studied direct, contrapositive, and contradiction arguments, as well as arguments which involve cases. The use of cases is really a subcategory of the other three, since you might use cases within any of a direct, contradiction, or contrapositive argument. For example, suppose you want to write a direct proof of a statement of the form $(p \vee q) \to r$. Here you would assume $p \vee q$, and then have p and q as your two cases. Or, perhaps you use a contrapositive argument to prove a statement of the form $p \to (\neg q \wedge \neg r)$. Here you could assume $q \vee r$, and use q and r as your cases.

The following table details what you should assume and what you need to conclude if you are proving a statement of the form $p \to q$ using either a direct, contrapositive, or contradiction argument:

Type of Argument	What you assume	What you must show
Direct	p	q
Contrapositive	$\neg q$	$\neg p$
Contradiction	$p \wedge \neg q$	This is impossible

In Class Activity: For this activity we ask that you consider ten proposed proofs of the statement: If $x + 15$ is even, then x is odd. For each of the proposed proofs, determine if it is an acceptable argument.

If it is an acceptable argument, answer all of the following:

- What type of logical argument did the author use?
- How well written was the proof?
- Was it easy to follow? Why or why not?
- Can you think of some specific details which would make it more clear? If so, what are they?

If it is not an acceptable argument, identify all the major problems that you find with the proof.

Proposed Proof 1. Suppose x is even.
Then $x = 2k$ for some integer k and so
$x + 15 = 2k + 15 = 2(k + 7) + 1$.
Since $k + 7$ is an integer, we see that $x + 15$ is odd.
Therefore if $x + 15$ is even, then x is odd. ∎

Proposed Proof 2. Suppose $x + 15$ is even and x is even.
Then $x + 15 = 2k$ and $x = 2q$ where k and q are integers.
Now note that $2q + 15 = 2k$ and so $15 = 2(k - q)$.
Therefore 15 is even.
But this is not true, thus if $x + 15$ is even, then x is odd. ∎

Proposed Proof 3. Suppose $x = 2k + 1$ where k is an integer.
Then $x + 15 = (2k + 1) + 15 = 2k + 16 = 2(k + 8)$.
The integers are closed under addition and so
$k + 8$ is an integer, say $k + 8 = q$.
Then we have $x + 15 = 2q$ and so $x + 15$ is even. ∎

Proposed Proof 4. Suppose $x + 15$ is even.
Then $x + 15 = 2k$ for some integer k.
So $x = 2k - 15 = 2k - 16 + 1 = 2(k - 8) + 1$.
And as $k - 8$ is an integer, x is odd. ∎

Proposed Proof 5. Suppose that x is even.
Then $x = 2k$, where k is an integer.
Consider that $x + 15 = 2k + 15 > 2k + 1$.
Therefore the statement is true. ∎

Proposed Proof 6. Suppose x is even.
Then $x = 2k$ where k is some integer.
Thus $x + 15 = 2k + 15$ which is odd.
Therefore $x + 15$ is even by proof by contradiction. ∎

Proposed Proof 7. Assume that if x is even,
then $x + 15$ is odd by the contrapositive.
So if x is even, then for some integer k, $x = 2k$.
Now $x + 15 = 2k + 15$.
This can also be written as $2k + 14 + 1 = 2(k + 7) + 1$.
Since $x + 15$ is odd, by contradiction we proved that
if $x + 15$ is even then x is odd. ∎

Proposed Proof 8. Suppose $x = 2k + 1$.
Then $x + 15 = (2k + 1) + 15 = 2k + 16 = 2(k + 8)$.
Therefore since $x + 15 = 2(k + 8)$
we conclude that $x + 15$ is even if x is odd. ∎

Proposed Proof 9. Assume x is even.
So $x = 2k$ where k is an integer.
Then $x + 15 = 2k + 15 = \frac{2}{15}k + 1 = 2q + 1$ where q is an integer.
Therefore by way of the contrapositive,
we have proven that if $x + 15$ is even, then x is odd. ∎

Proposed Proof 10. Suppose $x + 15 = 2k$ and $x = 2q + 1$,
where k and q are integers.
Now suppose that $2q + 1 + 15 = 2q + 16 = 2(q + 8)$.
Since q is an integer, we conclude that $2(q + 8)$ is even.
Therefore if $x + 15$ is even, then x is odd. ∎

Homework 2.7:

1. Critique the proposed proof of the statement provided. If the proof is correct, so state, noting which type of argument the author used. If it is incorrect, state one major problem with the proof.

   Statement. $\exists x \in \mathbb{R} \ni \forall y \in \mathbb{R}, y + x = 3$.
   Proposed Proof. Let y be any real number.
       Then note that if we let $x = 3 - y$, x is a real number.
       Furthermore, x is such that $y + x = y + (3 - y) = 3$.
       Therefore the statement is true. ∎

2. Critique the proposed proof of the statement provided. If the proof is correct, so state, noting which type of argument the author used. If it is incorrect, state one major problem with the proof.

   Statement. If x is irrational, then $x - 8$ is irrational.
   Proposed Proof. Suppose x is irrational and $x - 8$ is rational.
       Then $x - 8 = \frac{k}{r}$, where k and r are some integers, $r \neq 0$.
       Then $x = \frac{k}{r} + 8$, and so $x = \frac{k+8r}{r}$.
       Now since k and r are integers, $k + 8r$ is an integer and
       hence, x can be written in the form of an integer divided by an integer.
       But this is a contradiction since we assumed x is irrational.
       Therefore, the original statement is true. ∎

3. Prove the following using any type of appropriate argument(s):

   Let $x, y \in \mathbb{R}$. Prove $(x - y)^5 + (x - y)^3 = 0$ iff $x = y$.

4. Critique the proposed proof of the statement provided below. If the proof is correct, so state, noting which type of argument the author used. If the proofs is incorrect, state one major problem with the proof.

   Statement. If m is an odd integer, then $m + 6$ is an odd integer.
   Proposed Proof. For $m + 6$ to be odd, there must exist an integer n
       such that $m + 6 = 2n + 1$.
       Subtracting 6 from both sides of this equation and regrouping terms
       shows us that $m = 2n + 1 - 6$.
       Hence, $m = 2(n - 3) + 1$ and since we know n is an integer,
       we also know that $n - 3$ is an integer. Therefore, m is odd. ∎

5. Critique the proposed proof of the given statement. If it is correct, so state, noting which type of argument the author used. If the proofs is incorrect, state one major problem with the proof.

   Statement. $\exists x \in \mathbb{Z}, \exists y \in \mathbb{Z} \ni x + y = 1$.
   Proof. Suppose the statement is not true.
       Then there is an integer x such that for all integers y, $x + y \neq 1$.
       But if we let $x = 5$, then $y = -4$ is an integer for which $x + y = 1$.
       Therefore the negation is false, so the original statement is true. ∎

6. Prove that $\forall x \in \mathbb{R}, \frac{1}{x} \neq 0$.

7. Disprove that $\forall x, k \in \mathbb{Z}, x^k \in \mathbb{Z}$.

8. Prove that $\exists x \in \mathbb{Z} \ni x^2 + x = 2$.

9. Disprove that $\exists x \in \mathbb{Z} \ni x^2 + x = 1$.

10. Prove that $\forall m \in \mathbb{Z}, \exists n \in \mathbb{Z} \ni m > n$.

11. Prove that for all real numbers x and y, if $x \neq y, x > 0$, and $y > 0$, then $\frac{x}{y} + \frac{y}{x} > 2$.

12. Let $a, b, c \in \mathbb{Z}, b \neq 0$. Prove that if b divides a and b divides c, then b divides $3a + 2c$.

13. Let $x, y \in \mathbb{Z}$. Prove that if $x + y$ is odd and y is even, then x is odd.

14. Let $a, k, y \in \mathbb{Z}$ and suppose that $a = 5k + y$. Prove that if $5 \nmid y$, then $5 \nmid a$.

15. Let $a, b, c, d \in \mathbb{R}$. Prove that if $a > b > 0$ and $c > d > 0$, then $ac > bd$.

16. Prove the following: For every integer x, there is an integer y such that $3x + y = 6$.

17. Prove that if a is an integer, then $a(a^2 + 2)$ is divisible by 3.

18. Prove that $\exists x \in \mathbb{R} \ni \forall y \in \mathbb{R}, \frac{y}{x} \in \mathbb{R}$.

19. Let a, b, and c be nonzero integers. Prove that if $cx^2 + bx + a = 0$ has no rational solution, then $ax^2 + bx + c = 0$ has no rational solution. i.e. If $\nexists x \in \mathbb{Q} \ni cx^2 + bx + a = 0$, then $\nexists y \in \mathbb{Q} \ni ay^2 + by + c = 0$.

20. Prove that $\forall x \in \mathbb{Z}, \exists y \in \mathbb{Q} \ni 2x + 3y = 1$.

21. Prove that $\forall x \in \mathbb{R}, \exists y \in \mathbb{R} \ni |x + y| > 1$.

22. Disprove the following: The irrational numbers are closed under multiplication.

23. Disprove the following: The product of a rational number and an irrational number is an irrational number.

24. Disprove the following: If $x, y \in \mathbb{R}^+$ such that $\sqrt{\frac{x+y}{2}} \neq \frac{x+y}{2}$, then $x \neq y$.

25. Let $x \in \mathbb{Z}$. Prove that if $x^2 \equiv 1 \bmod 2$, then $x^2 \equiv 1 \bmod 4$. **Hint:** First prove the following lemma: If x^2 is odd then x is odd.

2.8   Regular Induction

We begin this section with some notation. You may have had the opportunity to see summation notation in calculus. Basically, $\sum_{i=1}^{3}(2i+1)$ is a way to represent $(2\cdot1+1)+(2\cdot2+1)+(2\cdot3+1)=3+5+7$. The sum-ends, $i=1$ to 3, indicate that we should begin with the variable $i=1$ and then to increase the variable i, in integer increments, until we get to 3. The sigma sign, $\sum$, indicates that we are to add all these terms together. Another notation similar to this is product notation. In product notation, $\prod_{i=1}^{3}(2i+1)$ represents $(2\cdot1+1)(2\cdot2+1)(2\cdot3+1)=(3)(5)(7)$.

Now consider the following statement: $\forall n \in \mathbb{N}$, $\sum_{i=1}^{n}i=\frac{n(n+1)}{2}$. This is a general statement about all natural numbers. It asserts that no matter what natural number you use for your top sum-end, equality will hold. For example, if you know the statement provided is true then you certainly know that $\sum_{i=1}^{5}i=\frac{5(5+1)}{2}$ is true since 5 is indeed a natural number.

Another way to write $\forall n \in \mathbb{N}$, $\sum_{i=1}^{n}i=\frac{n(n+1)}{2}$ is to write $\forall n \in \mathbb{N}$, $1+2+...+n=\frac{n(n+1)}{2}$. This later notion is also common (the use of ...) and so let us look also at some examples using this form.

> **In Class Activity 1:** Consider again the statement $\forall n \in \mathbb{N}, 1+2+...+n=\frac{n(n+1)}{2}$. What does this statement assert for $n=1$? What about for $n=5$?
>
> Next, consider the statement $\forall n \in \mathbb{N},\ 2^{0}+2^{1}+...+2^{n}=2^{n+1}-1$. What does this statement assert for $n=1$? What does it assert for $n=5$?

Induction is a type of argument in which you show that the statement in question is true for a starting value, and then go on to show that if it is true for some arbitrary integer k (where k is at least as big as your starting value), then the statement must be true for $k+1$. This process, often referred to as "the domino effect," shows the statement must be true for all integers greater than or equal to your starting value.

Before we demonstrate an induction proof we address some of the issues involved in induction proofs.

First note that such an argument will only prove the statement in question for integer values greater than or equal to some starting point. Hence induction cannot be used if you want to prove a statement to be true for all real numbers, or for all rational numbers, or even for all rational numbers greater than 6. Induction is only an applicable technique for proving statements which have a clear starting value and which state a claim about values which are greater than or equal to this starting value in integer increments.

> **In Class Activity 2**: Determine whether or not induction is an option as a method of proof for each of the following statements. *Note:* You don't have to see if the inductive argument actually works, just state whether or not the statement reflects the type of question for which induction would be an option.
>
>   1. If x is a real number and $x > 8$, then $x + 2 > 10$.
>
>   2. If x is an integer, then $x + 21$ is an integer.
>
>   3. If x is an integer and $x > 4$, then $x + 2 \geq -2$.
>
>   4. $\forall n \in \mathbb{N}$, $1 + 2 + ... (n - 1) + n = \frac{n(n+1)}{2}$.
>
>   5. For every positive real number $x > 1$, $x^2 > x$.

Another important issue regarding induction is that both components of the induction argument must be included. You must first show the statement in question is true for the initial value, and then go on to show that, if it is true for some arbitrary integer k (where k is at least as big as your starting value), then the statement must be true for $k + 1$. Consider the following statement along with its proposed proof.

Statement. For $n \in \mathbb{N}$, $1 + 3 + 5 + ... + (2n - 1) = n^2 + 3$.
Proposed Proof. Suppose that $k \in \mathbb{N}$ is such that
$$1 + 3 + 5 + ... + (2k - 1) = k^2 + 3.$$
Note that $1 + 3 + 5 + ... (2k - 1) + (2(k + 1) - 1)$
$= (k^2 + 3) + (2(k + 1) - 1)$
$= k^2 + 2k + 4$
$= (k + 1)^2 + 3.$
Therefore by the principle of mathematical induction, the statement is true for all $n \in \mathbb{N}$. ∎

Note that although this proposed proof correctly argues that if the statement is true for some integer k, then it is true for all integers greater than or equal to k, it fails to demonstrate that such an integer k exists in the first place. If it is fact true for a starting value, then by the above argument, it will be true from that point on (in integer increments). However, this statement is not true when $n = 1$ or when $n = 2$ or when $n = 3$. In fact, it is never true.

> **In Class Activity 3**: Consider the statement: For $n \in \mathbb{N}$, $1+3+5+ ... + (2n - 1) = n^2 + 3$. The statement looks as if it might be proven by induction however, it is not true. What does the statement assert when $n = 1$? What does the statement claim when $n = 2$? Should you be able to use induction to prove this statement is true? Why or why not?

In writing an induction proof, it is also a problem if you leave off the second portion — the argument that if the statement is true for some integer k, then it must be true for $k + 1$. The assumption that the statement is true for an arbitrary integer k (greater than or equal to your starting value) is often called the inductive assumption and while it may appear in this step that you are assuming that what you are trying to prove is true, we point out that you are only assuming it is true for some value, and calling that value k. This is a safe assumption since your first step was to demonstrate that this was the case. That is, in showing that the statement is true for your starting value, you have demonstrated that the statement is in fact true for some value.

We now present a valid induction proof:

Statement. If n is a natural number, then $1^2 + 2^2 + \ldots + n^2 = \frac{n(n+1)(2n+1)}{6}$.

Proof. Note that if $n = 1$ the statement is true since $1^2 = \frac{1(1+1)(2 \cdot 1 + 1)}{6}$.

i.e. $1 = \frac{1(2)(3)}{6}$.

Now suppose that k is a natural number for which the statement is true. That is, suppose that k is a natural number such that

$1^2 + 2^2 + \ldots + k^2 = \frac{k(k+1)(2k+1)}{6}$.

Note then that $1^2 + 2^2 + \ldots + k^2 + (k+1)^2 = \frac{k(k+1)(2k+1)}{6} + (k+1)^2$.

Also, note that this last expression can be simplified as follows:

$\frac{k(k+1)(2k+1)}{6} + (k+1)^2 = \frac{k(k+1)(2k+1) + 6(k+1)^2}{6} = \frac{(k+1)[k(2k+1) + 6(k+1)]}{6} = $

$\frac{(k+1)[2k^2 + 7k + 6]}{6} = \frac{(k+1)(k+2)(2k+3)}{6} = \frac{(k+1)[(k+1)+1][2(k+1)+1]}{6}$.

Thus, if the statement is true for some natural number k,

it also true for $k + 1$. Therefore, by the principle of mathematical induction, the statement is true for all natural numbers. ∎

The phrase "by the principle of mathematical induction" that appeared in the last sentence of our proof is often abbreviated by "BPMI" or "by the PMI."

Here is another statement which can be proven using induction, together with a valid proof of that statement:

Statement. If n is a natural number, then $3|(4^n - 1)$.

Proof. Note that when $n = 1$ the statement is true as $4^1 - 1 = 3$

and indeed $3|3$. Now suppose that k is a natural number such that $3|(4^k - 1)$.

Thus, $\exists q \in \mathbb{Z} \ni 3q = 4^k - 1$.

Note then that $4^{k+1} - 1$

$= 4 \cdot 4^k - 1$

$= (3 + 1) 4^k - 1$

$= 3 \cdot 4^k + (4^k - 1)$

$= 3 \cdot 4^k + 3q$

$= 3(4^k + q)$.

Now as k is an integer greater than or equal to 1, 4^k must be an integer (this is because the integers are closed under multiplication) and, as q is also an integer and we know that the integers are closed under addition, $4^k + q \in \mathbb{Z}$.

Thus, $3|(4^k - 1)$ and so BPMI, the statement is true for all $n \in \mathbb{N}$. ∎

In Class Activity 4: Consider the following proposed proof of the statement: For all integers $n \geq 0, 7^n = 1$.

Proof.[3] When $n = 0$ the statement is true since $7^0 = 1$.
Now suppose that $7^k = 1$ for some integer $k \geq 0$.
We know that $7^{k+1} = \frac{7^k \cdot 7^k}{7^{k-1}} = \frac{1 \cdot 1}{1} = 1$.
Therefore if the statement is true for some integer k, with $k \geq 0$, it is also true for $k + 1$.
Thus, by the principle of mathematical induction, the statement is true for all non-negative integers. ∎

Clearly there is a problem with the above proof since we know the statement is not true. In fact the statement is only true when $n = 0$. So, what is wrong with the logic used by the author?

Now that you have analyzed a few induction proof you are ready to try to write some on your own. To start you off on this we ask that you complete the following induction proof:

In Class Activity 5: Complete the proof of the statement: For any integer $n \geq 2$, $\left(\frac{2^2-1}{2^2}\right)\left(\frac{3^2-1}{3^2}\right)\cdots\left(\frac{n^2-1}{n^2}\right) = \frac{n+1}{2n}$.

Proof. Note that when $n = 2$ the statement is true
as $\left(\frac{2^2-1}{2^2}\right) = \frac{2+1}{2\cdot 2}$.
So suppose that k is an integer ≥ 2, for which we know
$\left(\frac{2^2-1}{2^2}\right)\left(\frac{3^2-1}{3^2}\right)\cdots\left(\frac{k^2-1}{k^2}\right) = \frac{k+1}{2k}$ Then note that
$\left(\frac{2^2-1}{2^2}\right)\left(\frac{3^2-1}{3^2}\right)\cdots\left(\frac{k^2-1}{k^2}\right)\left(\frac{(k+1)^2-1}{(k+1)^2}\right) = \left(\frac{k+1}{2k}\right)\left(\frac{(k+1)^2-1}{(k+1)^2}\right) = \cdots$
In order to correctly complete this proof, what is it that you need to show the expression on the right hand side of this last equation equals? Complete the proof by guiding your reader through the algebra that yields this result.

In Class Activity 6: Let $x \in \mathbb{Z}$. Prove that $\forall n \in \mathbb{N}$, $x^n \in \mathbb{Z}$.

---

[3]William Barnier and Norman Feldman, *Introduction to Advanced Mathematics* (Upper Saddle River, New Jersey: Prentice Hall, 1990) 56.

Homework 2.8:

For problems 1 & 2, evaluate the statements together with their proposed proofs. Determine if the statement is true and if so, if the proof is correct. If the statement is not true first disprove it and then determine what is wrong with the proof provided.

1. **Statement.** For any natural number n,
   $$3(1) + 3(2) + 3(3) + \dots + 3(n) = \frac{(3n^2+3n+2)}{2}.$$
   Proposed Proof.[4] Suppose that k is a natural number such that
   $3(1) + 3(2) + 3(3) + \dots + 3(k) = \frac{(3k^2+3k+2)}{2}$ and consider
   $3(1) + 3(2) + 3(3) + \dots + 3(k) + 3(k+1)$.
   By our induction assumption, this can be written as
   $\frac{(3k^2+3k+2)}{2} + 3(k+1)$.
   And this last expression can be written as
   $\frac{(3k^2+3k+2)}{2} + \frac{(6k+6)}{2} = \frac{(3k^2+3k+2+6k+6)}{2} = \frac{(3k^2+6k+3+3k+3+2)}{2}$
   $= \frac{(3(k^2+2k+1)+3(k+1)+2)}{2} = \frac{(3(k+1)^2+3(k+1)+2)}{2}$.
   Therefore, by the principle of mathematical induction, the statement is true for all natural numbers. ∎

2. **Statement.** For all natural numbers, $n^3 + 44n$ is divisible by 3.
   Proposed Proof.[5] Note that $1^3 + 44(1) = 45$ which is divisible by 3.
   So the statement is true for $n = 1$.
   Now suppose the statement is true for some natural number k.
   Then $k^3 + 44k$ is divisible by 3.
   Therefore $(k+1)^3 + 44(k+1)$ is divisible by 3.
   So BPMI, the statement is true for all natural numbers. ∎

3. Prove that for each natural number n, $1^3 + 2^3 + 3^3 + \dots + n^3 = \left[\frac{n(n+1)}{2}\right]^2$.

4. Consider each of the three proposed proofs for the result you proved in problem 3 above. Evaluate each as to whether or not it is correct and well written. If the proof is not correct, identify all major problems.

   Proposed Proof 1. Suppose that the equation is true for all natural numbers.
   Then $1^3 + 2^3 + 3^3 + \dots + k^3 = \left[\frac{k(k+1)}{2}\right]^2$ for $k \in \mathbb{N}$.
   Thus, $1^3 + 2^3 + 3^3 + \dots + k^3 + (k+1)^3 = \left[\frac{(k+1)(k+1+1)}{2}\right]^2$. ∎

   Proposed Proof 2. Note that if $n = 1$ the statement reads
   $1^3 = \left[\frac{1(1+1)}{2}\right]^2$ which is true.
   So suppose the statement is true for some natural number k.

---

[4]Peter Fletcher and C. Wayne Patty, *Foundations of Higher Mathematics* (Boston: PWS-Kent, 1991): 67-68.

[5]Douglas Smith, Maurice Eggen, and Richard St. Andre, *A Transition to Advanced Mathematics* (Pacific Grove, CA: Brooks Cole Thomson Learning 2001), 102.

Then $k^3 + (k+1)^3 = \left[\frac{k(k+1)}{2}\right]^2 + (k+1)^3 = \frac{k^2(k+1)^2}{4} + \frac{4(k+1)^3}{4}$

$= \frac{(k+1)^2(k^2+4(k+1))}{4} = \frac{(k+1)^2(k^2+4k+4)}{4} = \frac{(k+1)^2(k+2)^2}{4} = \left[\frac{(k+1)(k+1+1)}{2}\right]^2.$

Therefore by the PMI, the statement is true for all $n \in \mathbb{N}$. ∎

Proposed Proof 3. Note that if $n = 1$ the statement reads

$1^3 = \left[\frac{1(1+1)}{2}\right]^2$ which is true.

So suppose the statement is true for some natural number k.

That is, $1^3 + 2^3 + 3^3 + ... + k^3 = \left[\frac{k(k+1)}{2}\right]^2.$

Now consider $1^3 + 2^3 + 3^3 + ... + k^3 + (k+1)^3 = \left[\frac{(k+1)(k+1+1)}{2}\right]^2.$

We know that we can write this as $\left[\frac{k(k+1)}{2}\right]^2 + (k+1)^3.$

Furthermore $\left[\frac{k(k+1)}{2}\right]^2 + (k+1)^3 = \frac{k^2(k+1)^2}{4} + \frac{4(k+1)^3}{4}$

$= \frac{(k+1)^2(k^2+4(k+1))}{4} = \frac{(k+1)^2(k^2+4k+4)}{4} = \frac{(k+1)^2(k+2)^2}{4} = \left[\frac{(k+1)(k+1+1)}{2}\right]^2.$

Therefore BPMI, the statement is true for all natural numbers. ∎

For problems 5 through 9, determine whether or not induction is an option as a method of proof for the given statement. *Note*: You don't have to see if the inductive argument actually works, just state whether or not the statement reflects the type of question for which induction would be an option.

5. If n is any integer, then 8 divides $5^n + 2 \cdot 3^{n-1} + 1$.

6. If n is a positive integer, then $2^n > n^2$.

7. For every integer $n \geq 1$, $1(1!) + 2(2!) + ... + n(n!) = (n+1)! - 1$.

8. If n is an integer, then $n^2 + n + 1$ is odd.

9. If r is a real number, then $\sqrt{r^2} \geq r$.

Now we ask that you try to construct some induction proofs on your own:

10. Prove that for any integer $n, n \geq 1$, $n^3 + 5n + 6$ is divisible by 3.

11. Prove that for any natural number n, $2 + 5 + 8 + ... + (3n - 1) = \frac{n(3n+1)}{2}$.

12. Prove that for any natural number n, $5^n - 1$ is divisible by 4.

13. Prove that for every natural number n, $2^0 + 2^1 + ... + 2^n = 2^{n+1} - 1$.

14. Prove that for any natural number n, $7|(9^n - 2^n)$.

15. Prove that for each natural number n, $\frac{1}{1 \cdot 5} + \frac{1}{5 \cdot 9} + ... + \frac{1}{(4n-3)(4n+1)} = \frac{n}{4n+1}$.

16. Prove that for each natural number n, $21|(4^{n+1} + 5^{2n-1})$.

17. Prove that for each natural number n, $\frac{1}{2} + \frac{2}{2^2} + \frac{3}{2^3} + ... + \frac{n}{2^n} = 2 - \frac{n+2}{2^n}$.

18. Let $x \in \mathbb{R}$, $x \neq 1$. Prove that for any natural number n,

$1 + x + x^2 + ... + x^n = \frac{x^{n+1}-1}{x-1}$.

2.9   Induction with Inequalities

Next we look at using an induction argument to prove a statement which involves an inequality. When you need only show a quantity is greater than (or less than) another, you would think that it would be easier than showing it equal to something, however inequalities complicate matters a little since they open up the opportunity for some creativity. We explain this by providing such a statement along with a valid proof. In the proof we offer editorial notes to help guide the learner through the logic but for a final product (polished proof) the italicized notes should be omitted. That is, these notes should not be necessary for a person who understands how an inductive proof works, and when you write proofs, this is your target audience.

Statement.   For each natural number n, $n \geq 2$, $5^n > n + 4^n$.
Proof.   When $n = 2$ the statement is true as $25 = 5^2 > 2 + 4^2 = 18$.

This is all that is needed for our first step of the induction process. This last string would read "25 equals 5 squared which is greater than 2 plus 4 squared which equals 18." Thus, we have shown that in the case of $n = 2$, 5^n is indeed greater than $n + 4^n$.

So suppose that k is a natural number, $k \geq 2$, such that $5^k > k + 4^k$.

We have now completed the second step of an inductive argument which is to assume the statement is true for some arbitrary value k, which is at least as big as the largest value for which you evaluated the statement. Our next aim will be to show that the statement must be true for $k+1$. That is, we need to show that $5^{k+1} > (k+1) + 4^{k+1}$. Toward this end, we will begin by taking the quantity on one side, 5^{k+1}, and rewriting it into a form for which we can use our inductive hypothesis.

Now note that $5^{k+1} = 5 \cdot 5^k > 5 \left(k + 4^k \right)$

We have now applied our inductive assumption and, as our goal is to show that 5^{k+1} is greater than $(k+1) + 4^{k+1}$, we need to continue with our string above until we are able to determine that our quantity is greater than $(k+1) + 4^{k+1}$. Toward this end we work to rewrite $5 \left(k + 4^k \right)$ as close as possible to our desired form

$$= 5k + 5 \cdot 4^k$$
$$= 5k + (4+1) \cdot 4^k$$
$$= 5k + 4 \cdot 4^k + 4^k.$$

Now as $k \in \mathbb{N}$, $4^k > 0$, and so we know that
$$5k + 4 \cdot 4^k + 4^k > 5k + 4 \cdot 4^k = 5k + 4^{k+1}.$$

In the above we began by trying to rewrite the $5 \cdot 4^k$ term into a form which separated out a term of $4 \cdot 4^k$. Our reasoning should be clear. We need to get a term of 4^{k+1} which we know is $4 \cdot 4^k$. Our next goal is to somehow rework the remaining term $5k$ into the desired form of $k + 1$.

Furthermore, $5k + 4^{k+1} = (k + 4k) + 4^{k+1}$
and as $k \geq 2$, we know $4k \geq 8$ so certainly $4k > 1$.

Thus, $(k + 4k) + 4^{k+1} > (k + 1) + 4^{k+1}$.
We have shown that $5^{k+1} > (k + 1) + 4^{k+1}$, and so BPMI,
the statement is true for all natural numbers greater than or equal to 2. ∎

Recall that for a positive integer n, **n factorial,** denoted as $n!$, is defined by $n! = n \cdot (n - 1) \cdots 2 \cdot 1$. So, $3! = 3 \cdot 2 \cdot 1 = 6$ and $5! = 5 \cdot 4 \cdot 3 \cdot 2 \cdot 1 = 120$. Note also that we could write $5! = 5 \cdot 4!$, or $5! = 20 \cdot 3!$. This relationship allows us to write $(k + 1)!$ in terms of $k!$ and consequently enables induction to be a useful tool when it comes to proving relationships which involve factorial terms.

Consider the following statement and proof:

Statement. For all $n \in \mathbb{N}, n \geq 4, n! > 2^n$.
Proof. Note that if $n = 4$ the statement is true as
$4! = 24$, $2^4 = 16$, and $24 > 16$.
So suppose that k is a natural number, $k \geq 4$, such that $k! > 2^k$.
Note that $(k + 1)! = (k + 1) k!$ and as we know that $k! > 2^k$,
and that $(k + 1)$ is a positive integer, we can multiply both sides of our inequality by $(k + 1)$ to see that $(k + 1) \cdot k! > (k + 1) \cdot 2^k$.
Now note that as $k \geq 4$, we know that $k + 1 \geq 5$, and so certainly $(k + 1) > 2$.
Therefore we know that $(k + 1) \cdot 2^k > 2 \cdot 2^k = 2^{k+1}$.
Thus we see that $(k + 1)! > 2^{k+1}$ and so, BPMI,
the statement is true for all natural numbers n, where $n \geq 4$. ∎

Note again how the above proof involved an inequality which made it a little more difficult for us to determine how use the inductive assumption to generate our desired result for $k + 1$.

> **In Class Activity:** Consider the following statement together with a proposed proof. Determine what is wrong with the proposed proof. Once you have identified the error, determine whether or not the proof can be fixed. Either explain how, or why not.
>
> **Statement.** For all $n \in \mathbb{N}, n \geq 3, n! < 2^n$.
> **Proposed Proof.** If $n = 3$ the statement is true since $3! < 2^3$.
> Now suppose that $k \in \mathbb{N}$, $k \geq 3$, and that $k! < 2^k$.
> Note that $(k + 1)! = (k + 1) \cdot k! < (k + 1) 2^k < 2 \cdot 2^k = 2^{k+1}$.
> Thus BPMI the statement is true for all natural numbers. ∎

Homework 2.9:

1. Prove that for all integers $n \geq 6$, $2^{n+2} \leq n!$

2. Prove that for all integers $n \geq 2$, $2^n > 1 + n$.

3. Prove that for every integer n, where $n \geq 3$, $n^2 > 2n + 1$.

4. Prove that for all integers $n \geq 5$, $2^n > n^2$. *Hint:* You may want to invoke problem 3 above as a lemma for your proof of this problem.

5. Prove that for all natural numbers $n \geq 5$, $5n + 1 < 2^n$.

6. Prove that for all natural numbers n, $2^n + n \leq 3^n$.

7. Prove that for all integers n, with $n \geq 2$, $n^3 > 2n + 1$.

8. Prove that for all natural numbers $n \geq 5$, $(n + 1)! > 2^{n+3}$.

9. Prove that for all integers $n \geq 6$, $n! > n^3$. *Hint:* First prove the following lemma: If $n \in \mathbb{Z}, n \geq 6$ then $n^3 \geq (n + 1)^2$ and then proceed with your proof.

   In this last problem it is important to note that mathematicians do not know in advance of proving a statement that a lemma will be necessary. It is only when we work on a proof and, in the midst of our labor, realize that if we had another result things would work nicely, that we pull back and write a lemma. Lemmas provide us with a way to help our readers more easily follow our proof. If instead of writing a separate lemma here you had tried to include a proof of the second statement within your proof of the original statement, the result would be hard for a reader to follow. Sub-proofs within a proof seem tangential and make it harder for the reader to follow the overall logic of the original proof.

2.10 Recursion and Extended Induction

Consider the statement: $\forall n \in \mathbb{Z}, n \geq 3, \exists x, y \in \mathbb{Z} \ni n = 3x + 5y$. How might we proceed to prove such a statement? If we were going to attempt this proof as a direct/existence proof, we would begin by letting n be an appropriate integer and then trying to determine what form x and y need to take on in order for $3x + 5y$ to equal n. Unfortunately, choosing the form of x and y is not at all obvious. We also might try to use induction to prove this statement. In such a case we would begin by noting that if $n = 3$, the statement is true since we see that $x = 1$ and $y = 0$ yield the desired result. So we would next suppose that k is an integer greater than or equal to 3 for which the statement is true. That is, we know that $\exists x_1, y_1 \in \mathbb{Z} \ni n = 3x_1 + 5y_1$. Next we would need to determine integers x and y such that $k + 1 = 3x + 5y$. This is possible. That is, we could determine what integers x and y would yield $k + 1 = (3x_1 + 5y_1) + 1 = 3x + 5y$. What would they be? Once you've correctly determined this you could put together a nice induction proof of our statement. However, there is another form of induction which this question motivates.

Wouldn't it have been nice if instead of going from k to $k + 1$, we could have gone from k to $k + 3$? Then knowing that the statement was true for k, that is knowing that $\exists x_1, y_1 \in \mathbb{Z} \ni n = 3x_1 + 5y_1$, we could have noted that $k + 3 = (3x_1 + 5y_1) + 3 = 3(x_1 + 1) + 5y_1$ and since $x_1, y_1 \in \mathbb{Z}$, we know that $x_1 + 1, y_1 \in \mathbb{Z}$. Now this might seem like a big jump but it is a jump that would make perfectly good sense if we had started our initial induction step by not only showing that the statement was true for $n = 3$, but also for $n = 4$, and $n = 5$. Then the assumption that the statement is true for k together with proving it must be true for $k + 3$ would cover all integers greater than or equal to 3. Why? Because our initial step of showing the statement to be true for 3, together with our knowledge that if it's true for k, it's also true for $k + 3$, allows us to know it must be true for 3, 6, 9, ... Our initial step of showing the statement to be true for 4, together with our knowledge that if it's true for k, it's also true for $k + 3$, assures us that it must be true for 4, 7, 10, ... Finally, our initial step of showing the statement to be true for 5, coupled with our knowledge that if it's true for k, it's also true for $k + 3$, allows us to know it must be true for 5, 8, 11, ... All this together assures us that the statement must be true for all integers greater than or equal to 3.

This type of inductive argument in which you start off by showing the statement is true for several values in a row (not just for one value), is called extended induction and it is a technique which you would use in any inductive situation in which you either want to make your inductive argument skip over some values (as was the case in our last example) or in which you want to assume that the statement is true for several values, not just for k (we saw an attempt to do this in In Class Activity 3 of Section 2.8). We'll look at some proofs for which you will need to use extended induction, but first, we introduce recursion.

A **recursive sequence** is a sequence whose terms are defined in terms of proceeding terms of the sequence. For example, the sequence 1, 2, 3, 4, 5, ... could be defined *recursively* as follows: Let $a_1 = 1$, and for $n \in \mathbb{N}$, $n > 1$ define $a_n = a_{n-1} + 1$. Note that we could also define this sequence *explicitly* by for $n \in \mathbb{N}$, define $a_n = n$.

One particularly well known (and quite intriguing) sequence is the Fibonacci sequence. This sequence begins with two 1's and, after that, terms are obtained by adding together the two previous terms in the sequence. Hence the sequence goes 1, 1, 2, 3, 5,... and, since terms are computed from other terms in the sequence, this sequence is typically defined recursively as follows: $f_1 = 1$, $f_2 = 1$ and for $n \in \mathbb{N}$, $f_{n+2} = f_{n+1} + f_n$. While it is not at all obvious, this sequence could also be defined explicitly as follows: For each $n \in \mathbb{N}$, $f_n = \frac{\left(\frac{1+\sqrt{5}}{2}\right)^n - \left(\frac{1-\sqrt{5}}{2}\right)^n}{\sqrt{5}}$.

As another example, consider the sequence which is defined recursively by $b_1 = 16$ and, for $n \in \mathbb{N}$, $b_{n+1} = \frac{1}{2}b_n$. Note that this sequence yields the terms: $16, 8, 4, 2, 1, \frac{1}{2}, \frac{1}{4}, \ldots$ and so this sequence could also be described explicitly as follows: For $n \in \mathbb{N}$, $b_n = \frac{16}{2^{n-1}}$. Suppose that we wanted to convince someone (maybe even ourselves) that what we have stated here is in fact true. That is, we wanted to prove that the explicit formula does indeed generate the same sequence as the recursive formula. Toward this end, we offer the following induction proof:

Statement.   Let $b_1 = 16$ and for $n \in \mathbb{N}$ let $b_{n+1} = \frac{1}{2}b_n$. For $n \in \mathbb{N}$, $b_n = \frac{16}{2^{n-1}}$.
Proof.   Note that if $n = 1$ the statement is true since it asserts that
$b_1 = \frac{16}{2^0} = 16$ which is indeed what the recursive formula yields for b_1.
Now suppose that k is a natural number, $k \geq 1$,
such that the statement is true for k.
That is, suppose that it is indeed true that for this k, $b_k = \frac{16}{2^{k-1}}$.
Note that from our recursive formula $b_{k+1} = \frac{1}{2}b_k$ and hence,
$b_{k+1} = \frac{1}{2}\frac{16}{2^{k-1}} = \frac{16}{2^k} = \frac{16}{2^{(k+1)-1}}$.
Thus BPMI the explicit formula works for all $n \in \mathbb{N}$. ∎

We now return to our discussion on the extended principle of mathematical induction. As we noted earlier sometimes when writing proofs, particularly proofs which involve recursive relations, it is desirable to know that the initial statement works not just for some value k, but for several values (perhaps k and $k - 1$, or perhaps k, $k - 1$, and $k - 2$, or perhaps k and $k - 3$). These assumptions are perfectly valid as long as in your initial step of verifying that the statement is true for the first value, that you verify the statement is true for enough initial values for all your assumptions. For example, if you want to assume the statement is true for k, $k - 1$, and $k - 2$, then you would have to show by hand that the statement is indeed true for three initial values in a row (if the statement was about natural numbers then your initial step would be to show the statement is true for 1, 2, and also for 3 instead of just showing it was true for 1).

We now demonstrate this process. Consider the following statement together with a valid extended induction proof of that statement:

Statement.   Let $a_1 = 1$, $a_2 = 3$, and for $n \in \mathbb{N}$, $n \geq 3$, let $a_n = 3a_{n-1} - 2a_{n-2}$. For $n \in \mathbb{N}$, $a_n = 2^n - 1$.

Proof. If $n = 1$ the statement is true since the explicit formula yields
$a_1 = 2^1 - 1 = 1$ which is the desired result.
   If $n = 2$ the statement is also true since the explicit formula yields
$a_2 = 2^2 - 1 = 3$ which, again, is the desired result.
   Now suppose that $k \in \mathbb{N}$, $k \geq 2$, and that the explicit formula
holds for k and $k - 1$.
   That is, we know that for this value of k, $a_k = 2^k - 1$, and $a_{k-1} = 2^{k-1} - 1$.
   Consider that $a_{k+1} = 3a_k - 2a_{k-1}$
$= 3(2^k - 1) - 2(2^{k-1} - 1)$
$= 3 \cdot 2^k - 3 - 2^k + 2$
$= (3 \cdot 2^k - 2^k) - 3 + 2$
$= 2 \cdot 2^k - 1 = 2^{k+1} - 1.$
   Thus the formula holds for $k + 1$ and so, by the extended PMI, the formula holds for all $n \in \mathbb{N}$. ∎

In Class Activity:

   1. Let $a_1 = 1$, and for each natural number $n > 1$,

      let $a_n = 2a_{n-1} + 1$.

      Prove that for all natural numbers n, $a_n = 2^n - 1$.

   2. Let $a_1 = 2$, $a_2 = 4$, and for each natural number $n > 2$,

      let $a_n = a_{n-1} - 3a_{n-2}$.

      Prove that for all natural numbers n, a_n is even.

   3. Let $a_1 = 3$, $a_2 = 2$, and for each natural number $n > 2$,

      let $a_n = 5a_{n-1} - 6a_{n-2}$.

      Prove that for all natural numbers n, $a_n = 7 \cdot 2^{n-1} - 4 \cdot 3^{n-1}$.

Homework 2.10:

1. Suppose you have been asked to prove a statement about all integers greater than or equal to 5. If you try to use induction and find that you need to be able to assume the statement is not only true for k, but also for $k - 3$. For what specific initial value(s) would you need to show the statement holds true?

2. Let $d_1 = 2$, and $d_n = \frac{d_{n-1}}{n}$ for all $n \in \mathbb{Z}$ with $n \geq 2$. Prove that for all $n \in \mathbb{N}$, $d_n = \frac{2}{n!}$.

3. Let $a_1 = 2$, $a_2 = 1$, and for each natural number $n > 2$, let $a_n = 2a_{n-1} - a_{n-2}$. Prove that for all natural numbers n, $a_n = 3 - n$.

4. Consider the sequence $a_1 = 1$, $a_2 = 5$, and for $n \geq 3$, $a_n = 5a_{n-1} - 6a_{n-2}$. Prove that for all natural numbers n, $a_n = 3^n - 2^n$.

5. Prove that any natural number can be written as $4x + 5y$ for some integers x, and y. i.e. $\forall n \in \mathbb{N}, \exists x, y \in \mathbb{Z} \ni n = 4x + 5y$.

6. Let $a_1 = 2$, $a_2 = 4$, and for $n \geq 3$, $a_n = 5a_{n-1} - 6a_{n-2}$. Prove that for every natural number n, $a_n = 2^n$.

7. Let $a_1 = a_2 = 1$, and for each natural number $n > 2$, let $a_n = a_{n-1} + 3a_{n-2}$. Prove that for each natural number $n > 3$, $a_n < 3^{n-2}$.

8. Suppose that $h_0 = 1$, $h_1 = 2$, $h_2 = 3$, and for all integers $n \geq 3$, $h_n = h_{n-1} + h_{n-2} + h_{n-3}$. Prove that for all integers $n \geq 0$, $h_n \leq 3^n$.

9. Let $b_1 = 3$, $b_2 = 6$ and for $k \geq 3$, where k is an integer, let $b_k = b_{k-2} + b_{k-1}$. Prove that for all natural numbers n, b_n is divisible by 3.

10. Let $f_1 = 1$, $f_2 = 1$, and for $n \geq 3$, let $f_n = f_{n-1} + f_{n-2}$. Prove that for all $n \in \mathbb{N}$, $f_1^2 + f_2^2 + \ldots + f_n^2 = f_n f_{n+1}$

11. Let $f(x) = \ln x$. Prove that for all $n \in \mathbb{N}$, $f^n(x) = \frac{(-1)^{n+1}(n-1)!}{x^n}$, where $f^n(x)$ is the nth derivative of $f(x)$. Note that $0! = 1$.

12. Let $f_1 = 1$, $f_2 = 1$ and for $n \in \mathbb{N}$, $n \geq 3$, $f_n = f_{n-1} + f_{n-2}$. Prove that for all $n \in \mathbb{N}$, $f_n = \frac{\left(\frac{1+\sqrt{5}}{2}\right)^n - \left(\frac{1-\sqrt{5}}{2}\right)^n}{\sqrt{5}}$.

2.11 Uniqueness Proofs, the WOP, and a Proof of the Division Algorithm

We first introduced existence proofs in Section 2.2. We now extend the idea of proving that an element with a given property exists to proving that not only does such an element exist, but it is unique. That is, suppose that you were asked to prove that $\exists! x \in \mathbb{R} \ni \frac{x}{x+1} = 5$.

To prove that such a unique real number exists, you must first demonstrate that one does indeed exist. That is your proof would begin along the lines of the following:

Proof. Note that $\frac{-5}{4} \in \mathbb{R}$ and $\frac{\frac{-5}{4}}{\frac{-5}{4}+1} = \frac{-5}{-5+4} = 5$.

However this is only part of the proof. We also want to show that $\frac{-5}{4}$ is the only real number with this property. Toward this end, we will show that if two such real numbers exist, then they must in fact be equal. Hence demonstrating that the solution we have shown, $\frac{-5}{4}$, must in fact be the only one. Here is the complete proof:

Proof. Note that $\frac{-5}{4} \in \mathbb{R}$ and $\frac{\frac{-5}{4}}{\frac{-5}{4}+1} = \frac{-5}{-5+4} = 5$.

To see that $\frac{-5}{4}$ is unique, suppose that x_1 and x_2 are two real numbers such that $\frac{x_1}{x_1+1} = 5$ and $\frac{x_2}{x_2+1} = 5$.

Note then that $\frac{x_1}{x_1+1} = \frac{x_2}{x_2+1}$.

Hence, $x_1(x_2 + 1) = x_2(x_1 + 1)$.

That is, $x_1 x_2 + x_1 = x_1 x_2 + x_2$.

$\therefore x_1 = x_2$.

$\therefore \frac{-5}{4}$ is the only real number x with the property that $\frac{x}{x+1} = 5$. ∎

In Class Activity:

1. Prove that $\exists! x \in \mathbb{Q} \ni 2x - 1 = 7$.

2. Prove that $\forall x \in \mathbb{Z}, \exists! y \in \mathbb{Z} \ni x - y = 0$.

We now introduce a new definition, that of a "least element" of a set.

An element x is a **least element** of a set A iff $x \in A$ and $\forall y \in A$, $x \leq y$.

If $A = \{2, 3, 4\}$, 2 is a least element of A. However if $A = (-3, 0]$, then A does not contain a least element. Using this definition together with induction we can prove a very important result called the **Well Ordering Principle** (abbreviated as WOP). The WOP states that every non-empty set of non-negative integers contains a least element. For a proof of the WOP, consider the following:

Theorem. Every non-empty set of non-negative integers contains a least element.

Proof. Let A be a non-empty set of non-negative integers
and suppose that A does not have a least element.
Note that $0 \notin A$ for if it were, it would be the a least element in A.
Now suppose that for some non-negative integer k,
0,1,2,...,k are all elements which fail to be in A.
Note then that we know $k + 1 \notin A$, for if $k + 1 \in A$,
$k + 1$ would be a least element in A.
Hence BPMI all integers greater than or equal to 0 are not in A.
But A is a non-empty set of non-negative integers,
therefore we have a contradiction.
Thus, A must have a least element. ■

Now that we have the WOP at our disposal we are able to revisit something we began early in this chapter. Recall that the division algorithm asserts the following: Given any integers a, and b with $b > 0$, there exists unique integers q and r with $0 \leq r < b$ such that $a = bq + r$.

We are now equipped to understand the following valid proof of this important result:

Theorem. $\forall a, b \in \mathbb{Z}, b > 0, \exists! q, r \in \mathbb{Z} \ni a = bq + r$.

Proof. Let $a, b \in \mathbb{Z}$ with $b > 0$ and consider the set
$$S = \{a - bz \mid z \in \mathbb{Z} \text{ and } a - bz \geq 0\}.$$
We know that S contains some elements for if $a \geq 0$,
choosing $z = 0$ yields $a \in S$, and if $a < 0$,
choosing $z = a - 1$, yields $a - b(a - 1) = a - ba + b = a(1 - b) + b$.
Now as $b \geq 1$, $1 - b \leq 0$. This means that $a(1 - b)$ is nonnegative,
therefore $a(1 - b) + b$ is positive, and so is an element of S.
Therefore, S is a non-empty set of non-negative integers.
Hence, by the WOP, S contains a least element, call it r.
Note that $r = a - bq$ for some $q \in \mathbb{Z}$, and we know $r \geq 0$ (as $r \in S$).
Also note that $r < b$ for if it were the case that $r \geq b$ then
$a - bq \geq b$ and so $a - b(q + 1) \geq 0$.
This means that $a - b(q + 1) \in S$, and since r is the least element of S,
we know that $r \leq a - b(q + 1)$.
But this means that $a - bq \leq a - b(q + 1)$ which simplifies to $0 \leq -b$
or $b \leq 0$, which is impossible.
Thus $0 \leq r < b$.
Hence we have shown that for $a, b \in \mathbb{Z}$ with $b > 0$,
$\exists q, r \in \mathbb{Z}$ with $0 \leq r < b$ such that $a = bq + r$.
We now need only show that q and r are unique.
Toward this end suppose not.
That is, suppose that $\exists q', r' \in \mathbb{Z}$ with $0 \leq r' < b$ where
$q' \neq q$ or $r' \neq r$ but $a = bq + r$ and $a = bq' + r'$.
Then we know that $bq + r = bq' + r'$ and so $b(q' - q) = r - r'$.
Therefore $b \mid (r - r')$.

But as $0 \le r < b$ and $0 \le r' < b$, we know that $-b < r - r' < b$.
Hence the only way that it is possible that $b|(r - r')$ is if $r - r' = 0$.
Thus $r = r'$. But this means that $bq = bq'$ and
since $b \ne 0$, dividing both sides of this equation by b yields that $q = q'$.
Hence, q and r are indeed unique. ∎

Homework 2.11:

1. Prove that $\exists! x \in \mathbb{Z} \ni 2x - 5 = 9$.

2. Prove that if $a = 7$ and $b = 2$, there exist unique integers q and r with $0 < r < b$ such that $a = bq + r$. *Note*: We proved this in general when we proved the division algorithm but here we ask that you prove it just for this special case.

3. Prove that $\exists! x \in \mathbb{Z} \ni 2x^2 - 3x = 2$.

4. Let $f(x) = 3x + 1$ and $g(x) = 6x + 5$. Prove that $\exists! x \in \mathbb{R}$ such that $f(x) = g(x)$.

5. Let $y \in \mathbb{Z}$, $y \neq 0$. Prove that $\exists! x \in \mathbb{Z}$ such that $\frac{x}{y} = 2$.

6. Prove that $\forall x \in \mathbb{R}^+, \exists! y \in \mathbb{R}^+ \ni xy = 1$.

7. Prove that $\forall x \in \mathbb{Z}, \exists! y \in \mathbb{Z} \ni x + y = 0$.

For problems 8-18, determine if the given set has a least element. If so, what is that least element?

8. $\mathbb{Z}$

9. $\mathbb{Q}$

10. $\mathbb{N}$

11. $[3, 5)$

12. $(3, 5]$

13. $\{x \in \mathbb{Z} \mid x^2 - 1 = 0\}$

14. $\{x \in \mathbb{Z} \mid x \leq 3\}$

15. $\mathbb{R}^+$

16. $\mathbb{Z}^+$

17. $\mathbb{Z}^*$

18. $\mathbb{R}^*$

Chapter 3

Sets, Relations, and Functions

3.1 Sets

We have already defined a set as a collection of distinct elements and have referred to several sets of numbers including $\mathbb{R}$, $\mathbb{Z}$, $\mathbb{Q}$, and $\mathbb{N}$. Note however that a set need not consist of numbers. It is well within the definition of a set for a set to consist of any type of object. For example, if we define a set E so that $E = \{a, d, g, dog, 1, 2, 3\}$, E has 7 elements. Note that $dog \in E$ and $g \in E$, but $o \notin E$. As another example let us consider a set whose elements are sets. Let $B = \{\{1, 2, 3\}, \{1\}, \{2, 3\}\}$. B has three elements, each of which is a set. So, we could say that $\{2, 3\} \in B$ or $\{1, 2, 3\} \in B$ or $\{1\} \in B$, but $1 \notin B$ because 1 not a set, and all of the elements of B are sets.

We should also note that it is perfectly acceptable for a set to contain no elements. Such a set is called the empty set or the null set, and can be denoted by $\varnothing$ or $\{\}$. One should note however that $\{\varnothing\}$ is not an empty set, rather it is a set with one thing in it, and that one thing is the empty set.

Before we do too much with sets let's make sure that this idea of what it takes for something to be an element of a set is clear.

In Class Activity 1: Define sets A, B, C, D, E, F, and G as follows:

$A = \{1, 2, 3\}$

$B = \{\mathbb{Z}, \mathbb{R}, \mathbb{Q}\}$

$C = \{x \in \mathbb{R} \mid 1 < x \leq 10\}$

$D = \{x \in \mathbb{Z} \mid 1 \leq x \leq 10\}$

$E = \{x \in \mathbb{R} \mid x \notin \mathbb{Z} \text{ and } 1 < x \leq 6\}$

$F = \{a, d, \{1, 2, 3\}\}$

$G = \{1, 1, 1, 2, 3\}$

and determine whether each of the following are true or false:

___$1 \in A$	___$\mathbb{Z} \in A$	___$\{1\} \in A$
___$1 \in B$	___$\mathbb{Z} \in B$	___$\{1\} \in F$
___$1 \in C$	___$\mathbb{Z} \in D$	___$\{1, 2, 3\} \in F$
___$1 \in D$	___$5 \in B$	___$\{1, 2, 3\} \in G$
___$1 \in E$	___$5 \in C$	___$\{1, 2, 3\} \in A$
___$1 \in F$	___$5 \in D$	___$1 \in G$

We now define what it means for a set to be a subset of another set:

A set A is called a **subset** of a set B, denoted $A \subseteq B$, iff
$\forall x \in A, x \in B$. Equivalently, A is a subset of B iff $x \in A$ implies
$x \in B$.

Using the same sets as above, specifically, $A = \{1, 2, 3\}$, $D = \{x \in \mathbb{Z} \mid 1 \leq x \leq 10\}$, $F = \{a, d, \{1, 2, 3\}\}$, and $G = \{1, 1, 1, 2, 3\}$, note that $A \subseteq D$ since every element of A is an element of D. Also, $G \subseteq A$ as every element of G is an element of A. Furthermore $\{1, 2, 3\} \subseteq G$ but $\{1, 2, 3\} \not\subseteq F$. The reason that $\{1, 2, 3\}$ is not a subset of F is because every element of $\{1, 2, 3\}$ is not an element of F, for example $1 \notin F$.

In Class Activity 2: Let $A = \{1, \{2, 3\}, \{4\}\}$, and answer each of the following as true or false:

___$2 \in A$	___$1 \in A$	___$1 \subseteq A$
___$\{2\} \in A$	___$\{2, 3\} \in A$	___$\{1\} \subseteq A$
___$\varnothing \in A$	___$2 \subseteq A$	___$\{\{2, 3\}\} \subseteq A$
___$\{2\} \subseteq A$	___$\{2, 3\} \subseteq A$	___$\varnothing \subseteq A$
___$\{1, 4\} \subseteq A$	___$\{1, \{4\}\} \subseteq A$	

Now that we have established the notions of elements of sets and subsets of sets, we define a special set called the power set. Let A be a set. We define the **power set** of A, denoted $\wp(A)$, to be the set whose elements are all the possible subsets of A. That is, for a set A, $\wp(A) = \{B \mid B \subseteq A\}$. Note that for any set A, $\varnothing \in \wp(A)$ since $\varnothing \subseteq A$. As an example, suppose that $A = \{1, \{1\}, 2\}$. Here $\wp(A) = \{\varnothing, \{1\}, \{\{1\}\}, \{2\}, \{1, \{1\}\}, \{1, 2\}, \{\{1\}, 2\}, A\}$. Take a minute to make sure this is clear and then try to work the homework problems.

Homework 3.1:

For problems 1 through 8, let $A = \{1, \{b\}, \{1, b\}\}$ and for each problem place an '"$\in$", a "$\subseteq$", or the word "neither" or "both" in the blank to indicate under which conditions each of the following statements are true.

1. 1___A

2. $\{1\}$___A

3. $\{b\}$___A

4. $\{\{b\}\}$___A

5. $\{1, b\}$___A

6. $\{1, \{b\}\}$___A

7. $\{\varnothing\}$___A

8. $\varnothing$___A

For problems 9 through 13, let $D = \{1, -1\}$, and
$E = \{\{-1, 1\}, \{-\frac{1}{2}, \frac{1}{2}\}, \{-\frac{1}{3}, \frac{1}{3}\}\}$

9. $\wp(D) =$___

10. True or False:$D \in E$

11. T or F:$D \subseteq E$

12. T or F:$\{1\} \in D$

13. T or F: $\{1\} \subseteq D$

For problems 14- 21 let $A = \{1, 2, 3\}$, $C = \{\}$, $D = \{1, \{1\}, 2, \{3, 2\}, \{\varnothing\}\}$, and $E = \{1, 2\}$

14. T or F: $A \subseteq D$

15. T or F: $C \subseteq D$

16. T or F: $\{1\} \in D$

17. T or F: $\{1\} \subseteq D$

18. T or F: $\{\{1\}\} \subseteq D$

19. $\wp(E) =$___

20. T or F: $\{1\} \in \wp(D)$

21. T or F: $\{1\} \subseteq \wp(D)$

22. Determine $\wp(C)$.

23. Determine $\wp(\wp(\{1\}))$.

3.2   Set Operations

Now that (we hope) it is clear what it means for an element to be a member of a set and for a set to be a subset of a set, we define several set operations.

The **union** of sets A and B, denoted $A \cup B$, is the set whose members are elements of A or B. That is, $A \cup B = \{x \mid x \in A \text{ or } x \in B\}$.

The **intersection** of sets A and B, denoted $A \cap B$, is the set whose members are elements of both A and B. That is, $A \cap B = \{x \mid x \in A \text{ and } x \in B\}$.

The **difference** of sets A and B, denoted $A - B$, is the set whose elements are elements of A and not of B. That is, $A - B = \{x \mid x \in A \text{ and } x \notin B\}$.

The **complement** of a set A, denoted in this text by A^c, is the set of all elements which are not in A. It is important to note that if you define a set to contain elements *not in* another set you need to have an understood "universal set." The universal set is typically identified as "U". That is, if our universal set is $U = \{1, 2, 3\}$, and $A = \{1\}$, then $A^c = \{2, 3\}$, whereas if our universal set is the set of integers and $A = \{1\}$, then $A^c = \{x \in \mathbb{Z} \mid x \neq 1\}$.

In Class Activity: Let $U = \mathbb{R}$, $A = [3, 5)$, $B = \{1, 2, 3, 5, 7, 9\}$, $C = (-\infty, 3)$, and $D = \{-1, 3, 5\}$. Determine each of the following, trying to write your answer for each as concisely as possible:

1) $A \cup C$	2) $B \cup D$	3) $A \cap C$	4) $B \cap D$
5) $A \cap B$	6) $A - B$	7) $B - A$	8) $B - C$
9) C^c	10) $(A \cup C)^c$	11) $(A - B)^c$	12) $A \cup B$

Homework 3.2:

For problems 1 through 3 use the following sets: $U = \{x \in \mathbb{Z} \mid 0 \le x \le 9\}$, $A = \{1, 3, 5, 7, 9\}$, $B = \{0, 2, 4, 6, 8, 9\}$, $C = \{1, 2, 4, 5, 7, 8\}$.

1. $A - B =$

2. $A \cup C^c =$

3. $(A \cap B)^c =$

4. If F and G are sets, what does it mean to say that $x \notin F - G$? That is, unpack this definition as much as you can and explain in simple terms where such an element x would be.

For problems 5 through 8 let
$A = \{1, 2, 3\}$, $B = \{3, 4, 8\}$, $D = \{1, \{1\}, 2, \{3, 2\}, \{\varnothing\}\}$, and $E = \{1, 2\}$.

5. $A \cap B =$

6. $A - E =$

7. $(A \cap B) \cup D =$

8. If the universal set is the positive integers, then $A^c =$

For problems 9-12, let $U = \{x \in \mathbb{Z} \mid 2 \le x \le 12\}$, $A = \{5, 7, 9\}$, $B = \{2, 3, 4, 5\}$, and $C = \{12, 11, 9\}$.

9. $A \cap C =$

10. $A \cup (B \cap C) =$

11. $(A \cup B)^c =$

12. $B - A =$

3.3   Set Theory

We will now look at proving some relationships among sets. Recall that for sets A and B, A is a subset of B iff $x \in A$ implies $x \in B$. Hence, if you wanted to prove that for given sets A and B, $A \subseteq B$, you would want to prove that "$x \in A$ implies $x \in B$." That is, "if $x \in A$, then $x \in B$."

We've already written proofs of this nature (If p, then q), so it should make sense that if we wanted to prove that a set A was a subset of a set B that we would begin by letting x be an element of A and then explaining why this arbitrary x must also be an element of B. It's just as we did when we proved that "If x is an even integer, then x^2 is even", for there we began by supposing that x was an even integer and then, proved from this, that x^2 must also be even.

Hence, without further ado, we provide a sample proof that a set is a subset of another set:

Statement. For any sets A and B, $A \cap B \subseteq A$.
Proof. Let A and B be sets and let $x \in A \cap B$.
   As $x \in A \cap B$, we know that $x \in A$ and $x \in B$,
   so certainly $x \in A$.
   Thus $A \cap B \subseteq A$.   ∎

Here is another example:

Statement.   For sets A and B, if $A \subseteq B$, then $B^c \subseteq A^c$.
Proof. Let A and B be sets such that $A \subseteq B$ and suppose that $x \in B^c$.
   Then $x \notin B$.
   Now as $A \subseteq B$ we know that if $x \in A$, then $x \in B$,
   but more importantly for us, we know that the contrapositive
   of this is true. That is, we know that if $x \notin B$, then $x \notin A$.
   Thus, since we know $x \notin B$, we know that $x \notin A$.
   That is, $x \in A^c$.
   $\therefore B^c \subseteq A^c$.   ∎

What if you wanted to disprove that a set is a subset of another set? If you wanted to prove that for some sets A and B, $A \nsubseteq B$ you would want to prove that it was not the case that $x \in A$ implies $x \in B$. Hence we would want an example of an element x which was in A but not in B. Does this make sense? We are wanting to show that it is not the case that $\forall x \in A, x \in B$ and hence we must show that the negation is true. That is we must show that it *is* the case that $\exists x \in A \ni x \notin B$.

Two sets A and B are said to be equal, denoted $A = B$, iff $A \subseteq B$ and $B \subseteq A$. Proofs that two sets are equal are also fairly straightforward (logically speaking) in that there are two parts to the proof, showing that each set is indeed a subset of the other. As an example consider the following statement and proof:

Statement. Let A and B be sets whose elements belong to some universal set U. $(A \cap B)^c = A^c \cup B^c$.

Proof. Let A and B be sets whose elements belong to
some universal set U and let $x \in (A \cap B)^c$.
Then $x \in U$ but $x \notin (A \cap B)$.
Since $x \notin (A \cap B)$, we know $x \notin A$ or $x \notin B$.
If $x \notin A$, then $x \in A^c$ and, consequently, $x \in A^c \cup B^c$.
Otherwise $x \notin B$. In this case we know $x \in B^c$ and so $x \in A^c \cup B^c$.
Thus, we know $x \in A^c \cup B^c$ and so $(A \cap B)^c \subseteq A^c \cup B^c$.
Now suppose that $x \in A^c \cup B^c$. Then we know $x \in A^c$ or $x \in B^c$.
If $x \in A^c$ then $x \notin A$ and so $x \notin A \cap B$. Therefore $x \in (A \cap B)^c$.
Otherwise, $x \in B^c$ and in this case $x \notin B$,
which assures us that $x \notin A \cap B$.
Thus we know $x \in (A \cap B)^c$.
Therefore $A^c \cup B^c \subseteq (A \cap B)^c$ and so $(A \cap B)^c = A^c \cup B^c$. ∎

Now suppose that you wanted to prove that two specific sets, A and B, were not equal. How would you do this? You would want to show that either $A \not\subseteq B$ or $B \not\subseteq A$. That is either $\exists x \in A \ni x \notin B$, or $\exists x \in B \ni x \notin A$.

Constructing examples of sets which display certain attributes is not always quick and easy hence, mathematicians often use a graphical tool to help them develop such examples. As a motivational example, suppose I wanted to give you an example of sets A and B such that $A \cap B = A - B$, how might I construct such sets? Before we work too hard to construct such sets, we introduce the graphical tool. A **Venn diagram** is a graphical display of a relationship among sets. A Venn diagram can be drawn for any number of sets, but is most commonly helpful when one needs to demonstrate the relationship among 1, 2, or 3 sets, all of whose elements are in some universal set U.

If we have two sets A and B (as was the case in our motivational example above) we would draw two overlapping circles inside a rectangle. The circles represent the sets A and B, while the rectangle represents the universal set U. The sets A and B might have all elements in common, some elements in common, or no elements in common, pending on where within the diagram you place elements.

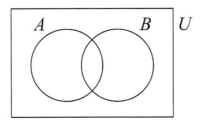

If we shade the area in the diagram that represents $A \cap B$ (with horizontal lines) as well as the area that represents $A - B$ (with slanted lines) we obtain the following:

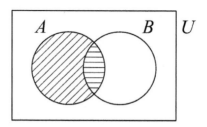

Now if we want an example of sets A and B for which $A \cap B = A - B$, we simply design our sets A and B so that exactly the same elements of A and of B fall in the area that is shaded by both diagrams. Since that area is empty, we simply want to make sure that our sets A and B are such that there are no elements in either $A \cap B$ or in $A - B$. So we want no elements to be in both A and B and no elements to be in $A - B$. This means we must set A to be the empty set and we can set B to be anything we like. So, an appropriate example would be to let $A = \varnothing$ and $B = \{2\}$. If, however, we wanted an example of sets A and B which demonstrated that $A \cap B \neq A - B$ we would simply need to make sure our sets A and B were designed so that some elements our sets were such that they fell in the area shaded by one of our diagrams but not the other. So an example of such sets would be $A = \{1\}$ $B = \{1\}$. Here $A \cap B = \{1\}$, but $A - B = \varnothing$.

We end this section with one last definition:

> Sets A and B are said to be **disjoint** iff they have no elements in common. That is, A and B are disjoint iff $A \cap B = \varnothing$.

Proofs that two sets are disjoint are often done by a contradiction argument. If you want to show that the intersection of two sets is empty, one way to do that is to assume that there is an element in both of these sets and then to show that this assumption leads to a contradiction. We provide an example to demonstrate this process:

> **Statement.** Let A and B be subsets of some universal set U. $(A \cap B^c) \cup (A^c \cap B)$ and $A \cap B$ are disjoint sets.
> **Proof.** Let A and B be subsets of some universal set U and suppose
> $\quad \exists\, x \in U$ such that $x \in (A \cap B^c) \cup (A^c \cap B)$ and $x \in A \cap B$.
> Since $x \in A \cap B$ we know that $x \in A$ and $x \in B$.
> So $x \notin A^c$ and $x \notin B^c$.
> As $x \notin A^c$, $x \notin (A^c \cap B)$ and as $x \notin B^c$, $x \notin (A \cap B^c)$.
> Thus $x \notin (A \cap B^c) \cup (A^c \cap B)$.
> But this is impossible as we chose x so that $x \in (A \cap B^c) \cup (A^c \cap B)$.
> Therefore no such element exists and hence
> $(A \cap B^c) \cup (A^c \cap B)$ and $A \cap B$ are disjoint sets. ∎

In Class Activity: Let A, B, and C be sets which are subsets of a universal set U.

1. Disprove the following: $A - B = B - A$.

2. Disprove the following: If $A \cup B = A \cup C$, then $B = C$.

3. Disprove the following: If $A \cup B = A \cup C$,
   then B and C are disjoint.

4. Prove that $(A \cup B)^c \subseteq A^c \cap B^c$.

5. Prove that $(A \cup B) - (A \cap B) \subseteq (A - B) \cup (B - A)$.

6. Prove that $A \cap B$ and $(A - B) \cup (B - A)$ are disjoint sets.

7. Prove that if $A \subseteq B$, then $A \subseteq B \cup C$.

8. Prove or disprove that $A \cap (B \cup C) = (A \cup B) \cap (A \cup C)$.

9. Prove that $\wp(A) \cup \wp(B) \subseteq \wp(A \cup B)$.

10. Prove that $\{(x, 2x - 2) \mid x \in \mathbb{R}\} = \{(t + 1, 2t) \mid t \in \mathbb{R}\}$.

Homework 3.3:

1. Draw a Venn diagram containing sets A and B, and shade the region which would correspond to $A^c \cap B^c$.

2. Draw a Venn diagram containing sets A, B, and C, and shade the region which would correspond to $(A \cup B) - C$.

3. Disprove that for any sets A, B, and C, if $A \cap B = A \cap C$, then $B = C$. *Note*: You want to provide an example of sets A, B, and C which meet the hypothesis but not the conclusion. To do this, we suggest you use a Venn diagram to help you think of appropriate sets, but the disproof should not be a diagram but rather the specific example that you develop. Remember, a disproof is like a proof, you do not show your scratch work, rather just your final, polished, argument.

4. Disprove that for any sets A, B, and C, $A - (B - C) \subseteq (A - B) - C$.

5. Prove that for any sets A, B, and C, $(A - B) \cap (A - C) \subseteq A - (B - C)$.

6. Let A, B, and C be sets. Prove that $B \cap (A \cup C) \subseteq A \cup (B \cap C)$.

7. Let A, B, C, and D be sets. Prove that if $A \cup B \subseteq C \cup D$, $A \cap B = \varnothing$, and $C \subseteq A$, then $B \subseteq D$.

8. Let A, B, and C be sets. Disprove that $A \cup (B \cap C) = (A \cup B) \cap C$.

9. Prove that for any sets A, B, and C, $A \cap (B \cup C) = (A \cap B) \cup (A \cap C)$.

10. Prove that for any sets A, B, and C, if $A \subseteq B$, then $A \cup (B \cap C) = B \cap (A \cup C)$.

11. Prove that for any sets A, B, and C, $(A - B) - C = A - (B \cup C)$.

12. Prove that for any sets A and B, $(A - B)^c = A^c \cup B$.

13. Prove that for any sets A, B, and C, $(A - B) \cap C$ and $(A - C) \cap B$ are disjoint.

14. Let A, B, C, and D be sets. Prove that if $A \subseteq B$, $C \subseteq D$, and $B \cap D = \varnothing$, then A and C are disjoint.

15. Prove that for any sets A, B, and C, if $B \cap C \subseteq A$, then $C - A$ and $B - A$ are disjoint.

3.4   Indexed Families of Sets

An **indexing set** is simply a set whose elements serve to index something. Indexing sets are often denoted by an upper case Greek lambda, "Λ" which correlates well with the word "label." Consider the following: Let $\Lambda = \{1, 2, 3\}$ and for each $i \in \Lambda$, define $A_i = \{0, i, i + 100\}$. Here Λ would be the indexing set as it provides the system of indices which will be used for the A_i's. Since our indexing set has three elements, the statement above defines three sets: $A_1 = \{0, 1, 101\}$, $A_2 = \{0, 2, 102\}$, and $A_3 = \{0, 3, 103\}$.

Once again recall summation notation (we referenced this earlier in the section on induction): $\sum_{i=1}^{3} 2i + 1$ is a way to represent $(2 \cdot 1 + 1) + (2 \cdot 2 + 1) + (2 \cdot 3 + 1) = 3 + 5 + 7$. The sum-ends ($i = 1$ to 3) tell us to begin with the variable $i = 1$ and then to increase the variable i in integer increments until we get to 3, while the sigma tells us to add all these terms together. Using a similar notation we can represent the union (or intersection) of an indexed family of sets. Another similar notation which you may have seen before is product notation. In product notation, $\prod_{i=1}^{3}(2i + 1)$ represents $(2 \cdot 1 + 1)(2 \cdot 2 + 1)(2 \cdot 3 + 1) = (3)(5)(7)$.

Similarly, $\bigcup_{i \in \Lambda} A_i$ represents the union of all A_i's where the i's are elements of the indexing set Λ. For our example provided above, this notation offers a condensed way to write $A_1 \cup A_2 \cup A_3$. Another option would be to write $\bigcup_{i=1}^{3} A_i$ but this last notation implies that our indexing set only consists of integers from 1 to 3 and, while that was the case in this example, it is quite possible that your indexing set might not consist of successive integers.

For example, suppose that your indexed family of sets is generated by the indexing set $\Lambda = \{1, 1.5, 2\}$ and for each $i \in \Lambda$, your sets are defined as $B_i = \{0, i, i + 100\}$. Note here that there are three sets in the indexed family. Here $\bigcup_{i \in \Lambda} B_i$ represents $B_1 \cup B_{1.5} \cup B_2$ but $\bigcup_{i=1}^{2} B_i$ would just represent $B_1 \cup B_2$.

In Class Activity:

For each $i \in \mathbb{N}$, define $A_i = \{1, 2, ..., i\}$.

1. Determine $\bigcup\limits_{i \in \mathbb{N}} A_i$.

2. Determine $\bigcap\limits_{i \in \mathbb{N}} A_i$.

Next, for each $i \in \mathbb{R}$, define $B_i = (i, i+1]$.

3. Determine $\bigcup\limits_{i \in \mathbb{R}} B_i$ as well as $\bigcap\limits_{i \in \mathbb{R}} B_i$.

Now, for each $i \in \mathbb{R}^+$, define $C_i = (-i^2, i^2]$.

4. Determine $\bigcup\limits_{i \in \mathbb{N}} C_i$ as well as $\bigcap\limits_{i \in \mathbb{N}} C_i$.

5. Determine $\bigcup\limits_{i \in \mathbb{R}^+} C_i$ as well as $\bigcap\limits_{i \in \mathbb{R}+} C_i$.

Let $A_\alpha = \{x \in \mathbb{Z} \mid x \neq \alpha\}$.

6. Determine $\bigcup\limits_{\alpha \in \mathbb{N}} A_\alpha$, as well as $\bigcap\limits_{\alpha \in \mathbb{N}} A_\alpha$.

Let $A_i = \{$all words in the English language which contain exactly i letters$\}$. Note that $A_0 = \varnothing$.

7. Determine $\bigcup\limits_{\alpha \in \mathbb{N}} A_\alpha$, as well as $\bigcap\limits_{\alpha \in \mathbb{N}} A_\alpha$.

Homework 3.4:

1. If $A_n = \left(\frac{-1}{n}, 2 - \frac{1}{n}\right)$ for every $n \in \mathbb{N}$, find $\bigcup_{n \in \mathbb{N}} A_n$.

2. If $A_n = \left(\frac{-1}{n}, 2 - \frac{1}{n}\right)$ for every $n \in \mathbb{N}$, find $\bigcap_{n \in \mathbb{N}} A_n$.

3. If $A_x = \{3, x\}$ for every $x \in \mathbb{R}$, find $\bigcup_{n \in \mathbb{N}} A_n$.

4. If $A_x = \{3, x\}$ for every $x \in \mathbb{R}$, find $\bigcap_{n \in \mathbb{N}} A_n$.

5. If $A_n = [\frac{-1}{n}, n]$, then $\bigcap_{n \in \mathbb{N}} A_n = $ ____.

6. If $A_n = \{-n, ..., -1, 0, 1, ..., n\}$, then $\bigcup_{n \in \mathbb{N}} A_n = $ ____.

7. Let $A_n = [2, 5 + \frac{1}{n})$. Find $\bigcap_{n \in \mathbb{N}} A_n$.

8. Note that if $x \in \bigcap_{i \in \Lambda} E_i$, then this means that $\forall i \in \Lambda, x \in E_i$. State clearly what it would mean if $x \notin \bigcap_{i \in \Lambda} E_i$.

9. State in terms of quantifiers what it means if $x \in \bigcup_{i \in \Lambda} F_i$ as well as what it means if $x \notin \bigcup_{i \in \Lambda} F_i$

10. Let B be a set and A_α, for $\alpha \in \Lambda$, an indexed family of sets. Prove that
$$B \cap \bigcup_{\alpha \in \Lambda} A_\alpha \subseteq \bigcup_{\alpha \in \Lambda} (B \cap A_\alpha).$$

11. Let Λ be an index set and A_α, for $\alpha \in \Lambda$, an indexed family of sets. Prove that $\bigcup_{\alpha \in \Lambda} A_\alpha$ and $\bigcap_{\alpha \in \Lambda} A_\alpha^c$ are disjoint.

12. Let Λ be an index set and A_α, for $\alpha \in \Lambda$, an indexed family of sets. Prove that
$$\left(\bigcup_{\alpha \in \Lambda} A_\alpha\right)^c = \bigcap_{\alpha \in \Lambda} A_\alpha^c.$$

3.5 Cartesian Products

We now consider a specific type of set, one whose elements consist of ordered n-tuples.

> The **Cartesian product** of sets A and B, denoted $A \times B$, is the set of all ordered pairs whose first coordinate is an element of A and whose second coordinate is an element of B. That is $A \times B = \{(x, y) \mid x \in A$ and $y \in B\}$.

As an example, consider $\mathbb{Z} \times \mathbb{R}$. Note that $(1, 2) \in \mathbb{Z} \times \mathbb{R}$, $(2, 1) \in \mathbb{Z} \times \mathbb{R}$, and $(1, 2.5) \in \mathbb{Z} \times \mathbb{R}$, but $(2.5, 1) \notin \mathbb{Z} \times \mathbb{R}$.

Note that $\mathbb{R} \times \mathbb{R}$ forms the 2-dimensional plane or the Cartesian coordinate system (and hence the reason for the term Cartesian product). $\mathbb{R} \times \mathbb{R}$ often gets denoted as $\mathbb{R}^2$ and so whenever you see $\mathbb{R}^2$ written, the author is referring to what you may have traditionally called the xy coordinate system. This idea of a Cartesian product can be extended to more than two coordinates. In the case of three coordinates, given sets A, B, and C, $A \times B \times C = \{(x, y, z) \mid x \in A$, $y \in B$, and $z \in C\}$. As you know, $\mathbb{R} \times \mathbb{R} \times \mathbb{R}$ forms what we sometimes call the 3-dimensional plane or the xyz-axis system and, as in the case for two dimensions, it is commonplace to denote $\mathbb{R} \times \mathbb{R} \times \mathbb{R}$, by $\mathbb{R}^3$. While we can't demonstrate with a graph higher dimensions than three, what is meant by $\mathbb{R}^n$ should be quite clear. It is the set of all n-tuples where each coordinate is a real number.

Homework 3.5:

1. If $A = \{1, 2, 3\}$ and $B = \{3, 4, 8\}$, then $A \times B$____

2. Construct a 2-dimensional plane and shade in the area which would correspond to the set $[1, 3) \times (2, 4]$, being careful to denote on your graph which values of your boundaries are, and are not, included in your answer.

3. Draw the graph of the Cartesian product $A \times B$ where $A = (1, 2]$ and $B = [2, 3)$.

4. Let $A = \{1, 2\}$, $B = \{0, 2\}$, and $C = \{a, b, c\}$. List all the elements of $C \times (A \cup B)$.

5. Let $A = [2, 3)$ and $B = (4, 5]$. Carefully sketch the region which contains the elements of $A \times B$. Do the same for $B \times A$. Note: Be sure to indicate which elements of the border of your region are included in this set.

6. Disprove that if A, B, C, and D are sets such that $A \subseteq B$ and $C \subseteq D$, then $B \times D \subseteq A \times C$.

7. Prove that for any sets A, B, and C, $A \times (B \cap C) = (A \times B) \cap (A \times C)$.

8. Let A, B, and C be sets. Prove that $A \times (B \cup C) \subseteq (A \times B) \cup (A \times C)$.

9. Let A and B be non-empty sets: Prove that if $A \neq B$, then $A \times B \neq B \times A$.

10. Let A and B be non-empty sets: Prove that if $A \times B \neq B \times A$, then $A \neq B$.

11. Comment to what extent the condition that the sets A and B be non-empty was necessary in problems 9 and 10.

12. Prove that for any sets A, B, and C, $A \times (B - C) \subseteq (A \times B) - (A \times C)$.

13. Prove that for any sets A, B, and C, $(A \cap B) \times C = (A \times C) \cap (B \times C)$.

14. Disprove that for any sets A, B, and C, $(A \cap B) \times C = (A \times B) \cap (A \times C)$.

3.6   Relations

A **relation** on a set A is an operation or rule that defines any subset of $A \times A$. As an example, if $A = \{1, 2\}$, then $\{(1, 2), (1, 1)\}$ is a relation on A. Another relation on A would be $\{(2, 2)\}$. Another would be $\{(1, 1), (2, 2), (1, 2), (2, 1)\}$. While these are all relations, mathematicians typically think of a relation as an operation or rule that assigns elements of a set to elements of that same set (and hence a rule that generates ordered pairs from $A \times A$). Consequently, a relation, typically denoted by "$\sim$" or by "R", is a rule or assignment that yields any subset of $A \times A$.

As an example consider the set of integers, $\mathbb{Z}$, and define a relation $\sim$ on $\mathbb{Z}$ so that $x \sim y$ iff $x - y = 3$. Note that $7 \sim 4$ since $7 - 4 = 3$. Therefore $(7, 4)$ is an element of the subset of $\mathbb{Z} \times \mathbb{Z}$ which is defined by $\sim$. Alternatively, you could say $(7, 4)$ is an element of the relation $\sim$. Also, note that $(4, 7)$ is not an element of the subset of $\mathbb{Z} \times \mathbb{Z}$ defined by $\sim$ as $4 - 7 \neq 3$.

In Class Activity 1:

1. List five elements of the relation $\sim$ on $\mathbb{Z}$ defined above.

2. For this same relation, list all integers x, such that $3 \sim x$.

3. Define a relation $\sim$ on $\mathbb{Z} \ni x \sim y$ iff $\frac{x}{y} \in \mathbb{Z}$.

   List 3 elements that are related to 10.

   i.e. Determine three values of $x \ni x \sim 10$.

4. Now define a relation $\sim$ on the set of integers by $x \sim y$

   iff $x - y = 0$. Describe the set that this relation defines.

We now introduce a few more definitions:

A relation $\sim$ on a set A is said to be **reflexive** iff $\forall x \in A, x \sim x$.

A relation $\sim$ on a set A is said to be **symmetric** iff $x \sim y$ implies that $y \sim x$.

A relation $\sim$ on a set A is said to be **transitive** iff $x \sim y$ and $y \sim z$ implies that $x \sim z$.

For a relation on a set to be reflexive, it must be the case that every element of the set is related to itself. Note that the relation $\sim$ defined in problem 3 above is not reflexive since $0 \in \mathbb{Z}$, but $0 \not\sim 0$. However, the relation $\sim$ in problem 4 above is reflexive since $\forall x \in \mathbb{Z}, x - x = 0$.

Also note that the relation $\sim$ defined in problem 3 above is not symmetric since $2 \sim 1$, but $1 \not\sim 2$. However, the relation defined in problem 4 is symmetric since if $x - y = 0$, then we also know that $y - x = 0$.

Finally, we note that the relation $\sim$ defined in problem 3 above is transitive since if $\frac{x}{y} \in \mathbb{Z}$, and $\frac{y}{z} \in \mathbb{Z}$, then since $\frac{x}{z} = \frac{x}{y} \cdot \frac{y}{z}$, we see that $\frac{x}{z}$ is the product of two integers and since $\mathbb{Z}$ is closed under multiplication, $\frac{x}{z} \in \mathbb{Z}$. Also, the relation $\sim$ defined in problem 4 is transitive since if $x - y = 0$, and $y - z = 0$, then adding these two equations together yields $x - z = 0$.

There is a special classification for relations which are reflexive, transitive, and symmetric, as was the relation on $\mathbb{Z}$ defined in problem 4.

> A relation $\sim$ on a set A is said to be an **equivalence relation** iff $\sim$ is reflexive, symmetric, and transitive.

If $\sim$ is an equivalence relation then the set of all elements which are related to an element x, is said to be the **equivalence class** of x, denoted $[x]$. That is, $[x] = \{y \in A \mid y \sim x\}$. For the equivalence relation defined in problem 4, $[8] = \{8\}$.

As another example of an equivalence relation, consider the following: Define a relation R on $\mathbb{Z}$ by aRb iff $a^2 - b^2$ is divisible by 3. We claim that R is an equivalence relation.

Proof. Let $x \in \mathbb{Z}$. Note that $x^2 - x^2 = 0$ which is divisible by 3.
Hence $\forall x \in \mathbb{Z}$, xRx and so R is reflexive.
Now suppose that $\exists x, y \in \mathbb{Z}$ such that xRy.
Then $\exists k \in \mathbb{Z} \ni x^2 - y^2 = 3k$.
But then note that $-k \in \mathbb{Z}$ and $y^2 - x^2 = 3(-k)$.
$\therefore y^2 - x^2$ is divisible by 3 and so yRx.
Hence, R is symmetric.
Finally suppose that $\exists x, y, z \in \mathbb{Z}$ such that xRy and yRz.
In this case we know $\exists a, b \in \mathbb{Z} \ni x^2 - y^2 = 3a$ and $y^2 - z^2 = 3b$.
Note that $x^2 - z^2 = (x^2 - y^2) + (y^2 - z^2) = 3a + 3b = 3(a + b)$.
And since we know $a + b \in \mathbb{Z}$, $x^2 - z^2$ is divisible by 3.
$\therefore xRz$ and so R is transitive.
As R is reflexive, symmetric, and transitive,
R forms an equivalence relation on $\mathbb{Z}$. ∎

For this equivalence relation note that $[1] = \{x \in \mathbb{Z} \mid xR1\}$. That is, $[1] = \{x \in \mathbb{Z} \mid \exists k \in \mathbb{Z} \ni x^2 - 1^2 = 3k\}$. That is, $[1] = \{x \in \mathbb{Z} \mid \exists k \in \mathbb{Z} \ni x^2 = 3k + 1\}$. Clearly $1 \in [1]$. Also, $2 \in [1]$, $3 \notin [1]$, and $4 \in [1]$. What else? Do you see a pattern? Determine all the distinct equivalence classes formed by R.

In Class Activity 2:

1. Define a relation $\sim$ on the set of integers by $m \sim n$ iff

   $(m - n)$ is divisible by 3.

   a. List three ordered pairs which would be in the subspace

   of $\mathbb{Z} \times \mathbb{Z}$ defined by $\sim$.

   b. Is $\sim$ transitive? Why or why not?

2. Define a relation R on the set of all people living in the

   world by aRb iff a lives within 100 miles of b.

   a. Is R symmetric? Why or why not?

   b. Is R transitive? Why or why not?

3. Define a relation $\sim$ on $\mathbb{R}$ by $x \sim y$ iff $y - x \in \mathbb{Z}$.
   Prove that $\sim$ is an equivalence relation.

4. For the equivalence relation defined in the problem 3 above, determine all elements in the equivalence class of 3. That is, determine $[3]$. Also determine $\left[\frac{1}{2}\right]$. What about $[\pi]$?

Homework 3.6:

1. Define a relation R on the set of integers by mRn iff $m|(n-2)$.

   (a) List three ordered pairs which would be in R.
   (b) Is R reflexive? Why or why not?
   (c) Is R symmetric? Why or why not?
   (d) Is R transitive? Why or why not?

2. Define a relation $\sim$ on the set of all people living in the world by $a \sim b$ iff a and b have ever lived on the same street.

   (a) Is $\sim$ reflexive? Why or why not? Prove your answer.
   (b) Is $\sim$ symmetric? Why or why not? Prove your answer.
   (c) Is $\sim$ transitive? Why or why not? Prove your answer.

3. Define a relation R on the set of real numbers by xRy iff x is approximately equal to y. Discuss the issue of transitivity regarding this relation.

4. Define a relation R on $\mathbb{Z}$ such that xRy iff $\frac{x}{y} \in \mathbb{Z}$. Prove that R is not symmetric.

5. Define a relation $\sim$ on $\mathbb{R}$ by $x \sim y$ iff $x \le y^2$. Prove that $\sim$ is not transitive.

6. Let $n \in \mathbb{N}$. Define a relation R on $\mathbb{Z}$ by aRb iff $a \equiv b \bmod n$. Prove that R is reflexive.

7. Let $n \in \mathbb{N}$. Define a relation R on $\mathbb{Z}$ by aRb iff $a \equiv b \bmod n$. Prove that R is symmetric.

8. Let $n \in \mathbb{N}$. Define a relation R on $\mathbb{Z}$ by aRb iff $a \equiv b \bmod n$. Prove that R is transitive.

9. Define a relation R on $\mathbb{Z}$ by xRy iff $3x + y$ is a multiple of 4. Prove that R is an equivalence relation. Determine $[0]$ and $[2]$.

10. Define a relation $\sim$ on $\mathbb{Z} - \{0\}$ by $a \sim b$ iff $a \cdot b > 0$. Prove that $\tilde{\ }$ is an equivalence relation and determine all (distinct) equivalence classes formed by this relation.

11. The relation $\sim$ defined on $\mathbb{R} - \{0\}$ by $a \sim b$ iff $\frac{a}{b} \in \mathbb{Q}$ forms an equivalence relation. Prove that $\left[\sqrt{3}\right] = \left[\sqrt{12}\right]$.

12. Let $A = \{(a,b) \in \mathbb{Z} \times \mathbb{Z} \mid b \neq 0\}$ and define a relation R on A such that $(x,y)R(w,z)$ iff $xz = yw$. Prove that R is transitive. Note: In your proof, point out what role the restriction that the second coordinate be non-zero plays.

13. Define a relation R on $\mathbb{R} \times (\mathbb{R} - \{0\})$ by $(a, b)R(c, d)$ iff $ad = cb$. Prove that R is an equivalence relation.

14. Suppose that $\sim$ is an equivalence relation on a non-empty set A.

 (a) The equivalence class of any element of A is always non-empty. Why?

 (b) We could define $[x]$ as $\{y \in A \mid y \sim x\}$ (as it is typically defined) or as $\{y \in A \mid x \sim y\}$. Why? Prove that these two sets are equal.

 (c) If $z \in [x]$, and $w \in [x]$, then $z \in [w]$. Why? Write a proof of this.

 (d) If $z \in [x]$, and $z \in [w]$, then $x \in [w]$. Why? Write a proof of this.

15. Prove the following: If $\sim$ is an equivalence relation on a set A and $x, y \in A$, then $[x] \cap [y] = \varnothing$ or $[x] = [y]$.

3.7   Functions

A **function** is a rule that assigns elements in a set A to elements in a set B, in such a way that each element of A is assigned to a unique element of the set B. The mathematical notation used to state that f is a rule that assigns elements in set A to elements of set B is "$f : A \to B$." Hence, $f : A \to B$ is a function iff $\forall x \in A, \exists! y \in B \ni f(x) = y$. In this case we say that the function f maps A to B and, within this context, the set A is called the **domain**, and the set B is called the **codomain**.

So, a function is a rule that assigns elements in the domain to elements in the codomain, and no element in the domain can be assigned to two different elements in the codomain (this is because a function must send each element in the domain to a *unique* element in the codomain).

Note that a function is similar to a relation but they are not the same. If the sets A and B were such that $A = B$, then any function from A to B would also form a relation, but the reverse is not the case. A function is much more restrictive than a relation. Moreover, functions need not be relations (when the domain and codomain are not equal) and relations need not be functions (since a relation can assign a single element of the domain to more than one element in the codomain).

Consider the function $f : \mathbb{Z} \to \mathbb{Q}$, where $f(x) = \frac{x}{2}$. Note that the domain of f is $\mathbb{Z}$ and the codomain of f is $\mathbb{Q}$. Note also that f assigns the integer 4 to the rational number 2. We say that 2 is the **image** of 4 under the function f. Or that 4 is a **preimage** of 2 under f. Note that every element in the domain must have an image, but every element of the codomain need not have a preimage. For example, $\frac{1}{3} \in \mathbb{Q}$ so this is an element of the codomain, but it has no preimage as $\nexists x \in \mathbb{Z} \ni f(x) = \frac{1}{3}$. If you wanted to prove this how would you do so? You would want to prove that no such element existed. One option for this would be to suppose that such an integer did exist and then to show that this leads to a contradiction. Let's try to do this. To be clear of what we're doing, we are trying to prove that there does not exist an integer $a \ni f(a) = \frac{1}{3}$ and we are going to do this via a contradiction argument.

Statement.   Let $f : \mathbb{Z} \to \mathbb{Q}$ be defined by $f(x) = \frac{x}{2}$. $\nexists a \in \mathbb{Z} \ni f(a) = \frac{1}{3}$.
Proof. Let $f : \mathbb{Z} \to \mathbb{Q}$ be defined by $f(x) = \frac{x}{2}$
and suppose $\exists a \in \mathbb{Z} \ni f(a) = \frac{1}{3}$.
Then we know that $f(a) = \frac{a}{2}$ and $f(a) = \frac{1}{3}$ thus $\frac{a}{2} = \frac{1}{3}$.
Therefore $a = \frac{2}{3}$.
But this is impossible since $a \in \mathbb{Z}$.
Therefore $\nexists a \in \mathbb{Z} \ni f(a) = \frac{1}{3}$. ∎

In Class Activity 1:

1. Reconsider our function $f : \mathbb{Z} \to \mathbb{Q}$ where $f(x) = \frac{x}{2}$.

   Since we have established that not all elements of the

   codomain have a preimage, determine all elements

   which do have a preimage.

2. Now consider the function $f : \mathbb{Q} \to \mathbb{Q}$ defined by $f(x) = \frac{x}{2}$.

What is different here? Do all elements in $\mathbb{Q}$ have a

preimage under this function? Why or why not?

For a function, the set of all elements in the codomain which have a preimage is called the **range**. Thus the range is always a subset of the codomain, and it may equal the codomain. Whenever the range and the codomain are the same, the function is said to be **onto** or **surjective**. An onto function is often called a surjection.

> Formally stated, a function $f : A \to B$ is said to be **onto** or **surjective** iff $\forall y \in B, \exists x \in A \ni f(x) = y$.

Also, we noted before that a function must take each element of the domain to a unique element of the codomain. So, if you defined a rule g such that $g(1) = 2$ and $g(1) = 3$, this would not represent a function (no matter what your domain and codomain was). But, it is perfectly acceptable that a function assign two different elements of the domain to the same element in the codomain. For example $f : \{1, 2\} \to \{3, 4\}$ where $f(x) = 3$ is a perfectly well defined function. Each element of the domain, $\{1, 2\}$, is assigned to a unique element in the codomain, $\{3, 4\}$. In this case the function is not onto since 4 is an element of the codomain which has no preimage.

> A function $f : A \to B$ is said to be **one-to-one** or **injective** iff $x \neq y$ implies that $f(x) \neq f(y)$.

A one-to-one (denoted as 1-1) function, is often called an injection. The function f, which we defined in the previous paragraph, is not 1-1 since there exists an $x \neq y$ for which $f(x) = f(y)$. That is, $1 \neq 2$ but $f(1) = f(2)$, thus f is not 1-1.

It should be clear that if you want prove a function f is 1-1 that you would want to use a contrapositive argument. That is, you would want to suppose that $\exists x, y$ in the domain such that $f(x) = f(y)$, and then to show that in fact $x = y$. We hope it is also clear that to prove a function f is onto you would need to construct an existence proof. That is, to prove that $f : A \to B$ is onto you would need to let $y \in B$ and then demonstrate an element $x \in A$ such that $f(x) = y$. This x that you demonstrate may or may not depend on y but it must be an element that you know exists in A.

As with sets, there is a helpful tool to help you think about how you might construct an example of a function with a given property (such as a function which is onto but not 1-1). The tool to which we are referring is a **mapping diagram** and mapping diagrams, like Venn diagrams, offer us a way to construct specific examples. A mapping diagram is particularly useful when your domain and codomain both contain a small number of elements. As an example consider the following:

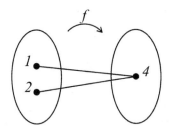

This display represents an assignment f, with domain $\{1,2\}$ and codomain $\{4\}$. By the arrows we see that $f(1) = 4$ and $f(2) = 4$. In this case, f represents a function since f assigns each element of the domain to a unique element of the codomain.

In Class Activity 2: Use the following mapping diagrams:

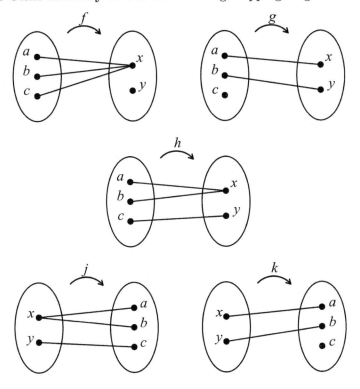

1. Which of the mapping diagrams given represent a function?
2. Of those which are functions, which are 1-1?
3. Of those which are functions, which are onto?

In Class Activity 3:

1. Let $f : \mathbb{Z} \to \mathbb{R}$ be defined by $f(x) = 2x - 5$.

   Prove that f is not onto.

2. Let $f : \mathbb{Q} \to \mathbb{Z}$ be defined by $f(q) = ab$, where $\frac{a}{b}$ is the
   reduced fraction form of q. Prove that f is onto.

3. Let $f : \mathbb{Z} \to \mathbb{Q}$ be defined by $f(z) = \frac{z}{z^2+1}$.

   Prove that f is not onto.

4. Let $f : \mathbb{Z} \to \mathbb{R}$ be defined by $f(x) = 2x - 5$.

   Prove that f is 1-1.

5. Let $f : \mathbb{R} \to \mathbb{R}$ be defined by $f(x) = x^2$.

   Prove that f is not 1-1.

Before we provide you with some practice problems to try on your own, we offer one last definition: A **bijection** is a function that is both 1-1 and onto.

We hope that it is now becoming clear to you that the basic proof argument forms we introduced in Chapter 2 are going to be forms you use in many mathematical settings. For each new class or topic you will be presented new definitions (such as we have done here with a function being 1-1 or onto). Those definitions will take on a form that you will then analyze and determine how best to write an argument in which you would prove or disprove the criteria. Consequently, from the form of the definition, you should be able to tell how best to prove or disprove whether something satisfies the necessary criteria.

Homework 3.7:

For problems 1 through 4, determine whether or not the described assignment represents a function. For any which fail to be a function be sure to explain why:

1. Let $f : \mathbb{Z} \to \mathbb{Z}$ be defined by $f(z) = \begin{cases} 1 \text{ if } z \text{ is divisible by } 2 \\ 2 \text{ if } z \text{ is divisible by } 3 \end{cases}$

2. Let $f : \mathbb{Z} \to \mathbb{Z}$ be defined by $f(z) = \begin{cases} 1 \text{ if } z \text{ is divisible by } 2 \\ 2 \text{ if } z \text{ is not divisible by } 2 \end{cases}$

3. Let $f : \mathbb{R} \to \mathbb{Z}$ be defined by $f(x) = 2x$.

4. Let $f : \mathbb{Z} \to \mathbb{R}$ be defined by $f(x) = 2x$.

5. Give an example of a function whose domain is the real numbers and whose codomain is the integers, or state why no such example can exist.

6. Define a function $f : \mathbb{Z} \times \mathbb{Z} \to \mathbb{Q}$ by $f((x, y)) = \frac{x+y}{3}$. List three elements in the pre-image of 2.

7. Consider the function $f : \mathbb{R} \to \mathbb{R}$ defined by $f(x) = \frac{2x-1}{3}$. Prove that f is onto.

8. Consider the function $f : \mathbb{Z} \to \mathbb{Q}$ defined by $f(x) = \frac{2x-1}{3}$. Prove that f is 1-1.

9. Disprove the following: The function $f : \mathbb{Z} \to \mathbb{R}$ defined by $f(x) = \frac{x}{2}$ is onto.

10. Consider the function $f : \mathbb{Z} \times \mathbb{Z} \to \mathbb{Z}$ defined by $f((a, b)) = a - b + 3$. Prove that f is onto.

11. If $f : Y \to Z$ is defined by $f(y) = z$, is 1-1 and onto, then the **inverse** of f, denoted f^{-1}, is the function which goes from Z to Y defined $f^{-1}(z) = y$.

    (a) Explain why it is the case that if f is not 1-1, then f^{-1} would not be function.

    (b) Explain why it is the case that if f is not onto, then f^{-1} would not be function.

12. Define $f : \mathbb{Z} \times \mathbb{Z} \to \mathbb{Z}$ by $f((x, y)) = x + y$. Prove that f is onto. Prove that f is not 1-1.

13. Let $f : \mathbb{Z} \times \mathbb{Z} \to \mathbb{Z}$ be defined by $f((a, b)) = ab$. Prove that f is onto (surjective).

14. Define $f : \mathbb{Z} \to \mathbb{Z} \times \mathbb{Z}$ by $f(x) = (2x, x^2)$. Prove that f is 1-1. Prove that f is not onto.

15. Define $f : \mathbb{R} \times \mathbb{R} \to \mathbb{R}$ by $f((a, b)) = ab$.

    (a) Prove that f is onto but not 1-1.
    (b) Determine the image of $(3, 2)$ under f.
    (c) Determine any and all elements which are in the preimage of 0 under f.

16. Let $f : \mathbb{Z} \to \mathbb{Z}$ be defined by $f(x) = 3x + 1$. Prove that f is 1-1 but not onto.

17. Let $f : \mathbb{Q} \to \mathbb{Q}$ be defined by $f(x) = 3x + 1$. Prove that f is a bijection.

18. Let $f : \mathbb{Z} \to \mathbb{R}$ be defined by $f(x) = 3x + 1$. Prove that f is not onto.

19. Let $f : \mathbb{Z} \times \mathbb{N} \to \mathbb{Q}$ be defined by $f((a, b)) = \frac{a-1}{b}$. Prove that f is onto but not 1-1.

20. Let $f : \mathbb{Z} \times \mathbb{N} \to \mathbb{R}$ be defined by $f((x, y)) = \frac{x}{y}$.

    (a) Determine any and all elements whose image under f is 0. That is, determine all elements in the preimage of 0.
    (b) Determine all elements whose image under f is 3.

21. Let $A = \{1, 2, 3, 4, 5\}$, and define a function $f : \wp(A) \to \mathbb{Z}$ as follows: For all sets X in $\wp(A)$, $f(X) = $ the number of elements in X (that is, the order of X).

    (a) What is the image of $\{1, 2, 3\}$ under f ?
    (b) What is in the pre-image of 4?
    (c) What is in the pre-image of 10?
    (d) Is f 1-1? Why or why not?
    (o) Is f onto? Why or why not?

22. Let $f : \mathbb{R} \to \mathbb{R}$ be defined by $f(x) = ax + b$ where a and b are both real valued constants (fixed real numbers), $a \neq 0$. Prove that f is a bijection, being clear to note if, and when, in your proof any of the given conditions on a and b are necessary.

3.8   Composition of Functions

Now suppose that f and g are functions such that $f : A \rightarrow B$ and $g : B \rightarrow C$. That is, suppose that f and g are functions such that the codomain of f is the domain of g. In this setting we can define the operation of composition. g composed with f, denoted "$g \circ f$," is a rule that assigns elements in A to elements in C by $(g \circ f)(x) = g(f(x))$. Does $g \circ f$ define a function? It should be clear that indeed $g \circ f : A \rightarrow C$. Is it the case that each element of A is assigned to only one element of C? Yes! Since f assigns each element of A to a unique B and g then assigns each element of B to a unique element in C, this new rule will also carry one element in A to only one element in C. Thus $g \circ f$ is indeed a function from A to C.

In Class Activity:

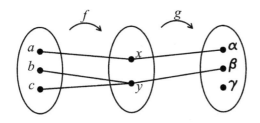

1. Using the mapping diagram above, determine if $g \circ f$ is 1-1 and/or onto.

2. Sketch a mapping diagram of functions f and g, where $f : A \rightarrow B$, $g : B \rightarrow C$, f is 1-1 and not onto, and g is 1-1 and onto.

3. Sketch a mapping diagram of functions f and g, where $f : A \rightarrow B$, $g : B \rightarrow C$, f is 1-1 and onto, and g is 1-1 and onto.

4. Sketch a mapping diagram of functions f and g, where $f : A \rightarrow B$, $g : B \rightarrow C$, f is not 1-1 and not onto, and g is 1-1 and onto.

Homework 3.8:

For problems 1-4, let $f : \mathbb{Z} \to \mathbb{Q}$ be defined by $f(z) = \frac{z}{2}$, and $g : \mathbb{Q} \to \mathbb{R}$ be defined by $g(q) = \pi \cdot q$.

1. Determine the domain of $g \circ f$.

2. Determine the value of $(g \circ f)(6)$.

3. Prove or disprove: $g \circ f$ is 1-1.

4. Prove or disprove: $g \circ f$ is onto.

For problems 5-16 Let f and g be functions such that $f : A \to B$ and $g : B \to C$.

5. Disprove the following: If f is 1-1, then $g \circ f$ is 1-1.

6. Disprove the following: If g is 1-1 then $g \circ f$ is 1-1.

7. Prove that if f and g are both 1-1, then $g \circ f$ is 1-1.

8. Disprove the following: If f is onto, then $g \circ f$ is onto.

9. Disprove the following: If g is onto, then $g \circ f$ is onto.

10. Prove that if f and g are both onto, then $g \circ f$ is onto.

11. Disprove the following: If $g \circ f$ is onto, then both g and f are onto.

12. Disprove the following: If $g \circ f$ is 1-1, then both g and f are 1-1.

13. Disprove the following: If f and $g \circ f$ are both 1-1, then g is 1-1.

14. Prove the following: If $g \circ f$ is 1-1, then f is 1-1.

15. Prove or disprove the following: If $g \circ f$ is onto, then g is onto.

16. Prove or disprove the following: If g and $g \circ f$ are both onto, then f is onto.

3.9 Cardinality

A set A is said to be **finite** iff A is empty or for some non-negative integer n, there exists a bijection $f : \{1, 2, 3, ..., n\} \to A$. Otherwise the set is said to be infinite. For example, $A = \{2, 3, 8\}$ is finite since if we define $f : \{0, 1, 2\} \to A$ by $f(0) = 2$, $f(1) = 3$, and $f(2) = 8$, f is 1-1 and onto.

Two sets, A and B, are said to have the same **cardinality** iff there exists a 1-1, onto function from A to B. The cardinality of a finite set is the number of elements in the set. Let E represent the set of all even integers. It is interesting to note that E and $\mathbb{Z}$ have the same cardinality since if we define $f : \mathbb{Z} \to E$ by $f(z) = 2z$, f is a bijection. Hence, E and $\mathbb{Z}$ have the same cardinality.

An infinite set B is said to be **countably infinite** iff there exists a 1-1, onto function $f : \mathbb{N} \to B$. If no such bijection exists then B is said to be **uncountably infinite**.

$\mathbb{Z}$ is countably infinite. To see this note that if we define $f : \mathbb{N} \to \mathbb{Z}$ by $f(x) = \begin{cases} \frac{x-1}{2} & \text{if } x \text{ is odd} \\ -\frac{x}{2} & \text{if } x \text{ is even} \end{cases}$,we see that indeed there exists a 1-1, onto function $f : \mathbb{N} \to \mathbb{Z}$. We should note to our reader that it is a worthwhile exercise to prove that this function f is indeed a bijection and we encourage you to think through how such a proof might go.

There is a very famous proof that the set of real numbers on the interval $(0, 1)$ is uncountably infinite. This proof was published by Georg Cantor in the late 1800s and it is often referred to as Cantor's Diagonalization Proof. The proof goes something like this:

Theorem. $(0, 1)$ is not a countably infinite set.

Proof. Suppose that $\mathbb{N}$ has the same cardinality as $(0, 1)$.

Then there is a function $f : \mathbb{N} \to (0, 1)$ which is 1-1 and onto.

We could then list all the values in $(0, 1)$ as the image of the function f as follows:

$f(1) = 0.d_{11}d_{12}d_{13}...$
$f(2) = 0.d_{21}d_{22}d_{23}...$
$f(3) = 0.d_{31}d_{32}d_{33}...$

.

.

.

Now since our assumption is that $\mathbb{N}$ has the same cardinality as $(0, 1)$, this list above of the elements of $(0, 1)$ should be exhaustive.

However, we claim that there is a number in $(0, 1)$ which cannot possibly be on this list (and hence it is not possible that a function which maps the natural numbers to $(0, 1)$ be onto).

To see this, consider the real number $x = 0.b_1b_2b_3...$

where, for each $i \in \mathbb{N}$, $b_i \neq d_{ii}$, $b_i \neq 0$, and $b_i \neq 9$.

This number x cannot be on the list as x differs from the ith number in the ith decimal place for every $i \in \mathbb{N}$.

Also, x is in the interval from 0 to 1.

Hence, $\mathbb{N}$ does not have the same cardinality as $(0, 1)$. ■

Corollary. $(0, 1)$ is uncountably infinite.

You may have wondered why, in the proof above, we insisted that $b_i \neq 9$. This is simply to avoid the number we construct being $x = .\overline{9}$ which in fact, equals 1.

Another non-trivial cardinality proof is one which shows that the set of rational numbers is countably infinite. Cantor developed a scheme much like the argument used above to prove that the set of positive rational numbers has the same cardinality as $\mathbb{N}$, and this idea can be extended to prove that $\mathbb{Q}$ is countably infinite.

Homework 3.9:

Prove each of the following:

1. The set of natural numbers is countably infinite.

2. $\mathbb{Z}^*$ is countably infinite.

3. The set of positive even integers is countably infinite.

4. The set of even integers has the same cardinality as the set of all multiples of 10.

5. The set of even integers is countably infinite.

6. $(-10, 10)$ has the same cardinality as $(0, 1)$.

7. $(0, 1)$ has the same cardinality as $(-10, 10)$.

8. Can you identify a relationship between the function you used in your proof of #6, and the function you used in proving #7?

9. $(0, 1)$ has the same cardinality as $(0, 2)$.

10. $(0, 2)$ has the same cardinality as $(-10, 10)$.

11. Can you identify a relationship between the functions you used in your proofs of #9 and #10 to that of the function you used in proving #7?

12. A set A has the same cardinality as itself. i.e. Cardinality is reflexive

13. Cardinality forms an equivalence relations on sets. *Note:* Be sure that you clearly define the functions which assure you that cardinality is reflexive, symmetric, and transitive.

14. Use the fact that $f : (\frac{-\pi}{2}, \frac{\pi}{2}) \to \mathbb{R}$ defined by $f(x) = \arctan x$ is 1-1 and onto, and the information you have learned from working the previous problems in this homework set, to prove that $(0, 1)$ has the same cardinality as $\mathbb{R}$.

Answers to Selected Homework

Chapter 1

1.1 Introduction and Terminology

1. False because $\sqrt{2} \in \mathbb{R}$ but $\sqrt{2} \notin \mathbb{Z}$.

5. True. Since all integers are real numbers it is impossible for there to be a non-real number which is also an integer.

6. $\frac{11}{9}$

1.2 Statements and Truth Tables

2.

p	q	$q \to p$	$p \wedge (q \to p)$
T	T	T	T
T	F	T	T
F	T	F	F
F	F	T	F

5. A statement of the form $p \wedge (q \to p)$ will be false whenever p is false.

1.3 Logical Equivalence and Logical Deductions

2.

p	q	$\neg(p \to q)$	$p \wedge \neg q$
T	T	F	F
T	F	T	T
F	T	F	F
F	F	F	F

6. They are not logically equivalent. If $x \notin S$ is true and $xy \notin T$ is false, then the first statement would be false while the second statement would be true.

12. Yes. From the third statement we get that "x even implies x^2 even" is true.

Using the first and second statement, we can see that the fourth statement is the contrapositive of "x^2 even implies x even."

Hence, we can conclude x is even iff x^2 is even.

14. A statement of this form would be true unless both p and q are false.

19. If for all i, $a_i < a_{i+1}$, then the sequence is increasing.

29. If somebody doesn't show me how to work number 3, then I am going to fail this test.

1.4 Contrapositive, Negation, and Converse

1. contrapositive: $\neg q \to \neg p$
   negation: $p \wedge \neg q$
   converse: $q \to p$

5. contrapositive: If f is continuous $x = a$, then f is differentiable at $x = a$.
   negation: f is not differentiable at $x = a$ and f is continuous at $x = a$.
   converse: If f is not continuous $x = a$, then f is not differentiable at $x = a$.

10. $(p \wedge \neg q) \vee (q \wedge \neg p)$

11. An integer x is not divisible by 6 iff x is not divisible by 2 or x is not divisible by 3.

1.5 Quantifiers

4. True, $4 \in \mathbb{Z}$ and $\sqrt{4} \in \mathbb{Z}$.

14. True. The statement makes a claim about the integers 1, 2, and 3.
    In the case of 1, $a = 1, b = c = 0$, produces the desired result.
    In the case of 2, $a = b = 1, c = 0$ works, and in the case of 3, $a = b = c = 1$ works.

18. False, there is not one real number which, when multiplied to every non-zero real number, will always equal 1.

27. $\exists x \in \mathbb{R} \ni x > 0$ and $\forall y \in \mathbb{R}, y \geq 0$ or $xy \leq 0$.

30. A sequence $\{a_n\}$ is not Cauchy iff there is an $\varepsilon > 0$, such that for every $N \in \mathbb{N}$, $m, n > N$, and $|a_m - a_n| \geq \varepsilon$.

Chapter 2

2.2 Existence Proofs and Counterexamples

1. Let $f(x) = 3x + 1$ and $g(x) = 6x + 5$. Note that $x = -\frac{4}{3}$ is a real number and $f(\frac{-4}{3}) = -3 = g(\frac{-4}{3})$.

4. Note that since $3(12) = (56 - 20)$, $12|(56 - 20)$. That is, $56 \equiv 20 \bmod 12$.

5. Note that $x = 8 \in \mathbb{Z}$ and $y = 4 \in \mathbb{Z}$ are such that $\frac{x}{y} = 2$.

6. Disproof of b: Note that if $x = y = 6, z = 1$, this statement is false because $6|(6 \cdot 1)$ but $6 \nmid 1$.
   Disproof of c: Note that $x = 6, y = 2, z = 3$ shows that statement is false because $6|(2 \cdot 3)$ but $6 \nmid 2$ and $6 \nmid 3$.

11. Note that $1, 2 \in \mathbb{Z}$ but $3(1) + 2(2) \neq 1$.

2.3 Direct Proofs

2. $x^2 + y = 13$ and $y \neq 4$.

5. **Proof.** Suppose $a \in \mathbb{Z}$.
   Note that $2a - 3 = 2a - (4 - 1) = 2a - 4 + 1$
   $= (2a - 4) + 1 = 2(a - 2) + 1$.
   And since we know $a - 2 \in \mathbb{Z}$, we know that $\exists k \in \mathbb{Z} \ni k = a - 2$.
   Hence, we know $2a - 3 = 2k + 1$ for some integer k.
   Therefore $2a - 3$ is odd. ∎

8. **Proof.** Let $a, b, c \in \mathbb{Z}$ and suppose that $a|b$.
   Then $\exists k \in \mathbb{Z} \ni b = ak$.
   Hence, $5b - 2a = 5(ak) - 2a = a(5k - 2)$.
   Furthermore, since $5, k, 2 \in \mathbb{Z}$, $5k - 2 \in \mathbb{Z}$.
   $\therefore a|(5b - 2a)$. ∎

10. **Proof.** Let $a, b \in \mathbb{Z}$ and $n \in \mathbb{Z}^+$ and suppose $a \equiv b \bmod n$.
    Then $\exists k \in \mathbb{Z} \ni nk = (a - b)$.
    Note then that $n(-k) = (b - a)$.
    And, since we know $-k \in \mathbb{Z}$, we know that $n|(b - a)$.
    $\therefore b \equiv a \bmod n$. ∎

2.4 Using Cases

4. There would be four cases for the value a: $a = 5k + 1$, $a = 5k + 2$, $a = 5k + 3$, $a = 5k + 4$. Then there would also be four cases for the value b: $b = 5q + 1$, $b = 5q + 2$, $b = 5q + 3$, $b = 5q + 4$. The actual cases in the proof would be every possible combination of a and b, that is, there are 16 total cases.

7. **Proof.** Let $x \in \mathbb{R}$. Then $x \geq 1$ or $x < 1$.
If $x \geq 1$, $|x - 1| = x - 1$ and $|1 - x| = -(1 - x) = x - 1$,
thus $|x - 1| = |1 - x|$.
If $x < 1$, $|x - 1| = -(x - 1) = 1 - x$ and $|1 - x| = 1 - x$.
And so again, $|x - 1| = |1 - x|$. ∎

9. **Proof.** Suppose x is an odd integer.
Then there is a $k \in \mathbb{Z} \ni x = 2k + 1$.
As $k \in \mathbb{Z}$, we know k is either even or odd.
Case 1: Suppose k is even.
Then $\exists q \in \mathbb{Z} \ni k = 2q$.
Note then that $x^2 = [2(2q) + 1]^2 = 16q^2 + 8q + 1$
$= 8(2q^2 + q) + 1$.
Since we know $(2q^2 + q) \in \mathbb{Z}$,
we see that $\exists m \in \mathbb{Z} \ni x^2 = 8m + 1$.
Case 2: Suppose k is odd.
In this case, $\exists q \in \mathbb{Z} \ni k = 2q + 1$.
Note that $x^2 = (2(2q + 1) + 1)^2$
$= (4q + 3)^2 = 16q^2 + 24q + 9$
$= 8(2q^2 + 3q + 1) + 1$.
And since we know $(2q^2 + 3q + 1) \in \mathbb{Z}$,
we see that x^2 is of the form $8m + 1$ for some integer m. ∎

2.5 Contrapositive Arguments

1. You would assume that $x = 3$.

4. **Proof.** Let $f(x) = 3x - 2$ for $x \in \mathbb{R}$, and suppose that $f(x) = f(y)$.
Then $3x - 2 = 3y - 2$.
Hence, $3x = 3y$.
$\therefore x = y$. ∎

7. **Proof.** Let $x \in \mathbb{R}$ and suppose that $\sqrt[3]{x}$ is rational.
Then $\exists a, b \in \mathbb{Z} \ni \sqrt[3]{x} = \frac{a}{b}$.
Note then that $x = \frac{a^3}{b^3}$.
Furthermore, since we know $a^3, b^3 \in \mathbb{Z}$, we see that x is rational. ∎

2.6 Contradiction Arguments

3. **Proof.** Let $x, y \in \mathbb{Z}$ and suppose that $8x + 2y = 3$.
Note then that $2(4x + y) = 3$.
But since we know $(4x + y) \in \mathbb{Z}$,
this means that $2|3$, which is impossible.
Thus, our assumption that $8x + 2y = 3$ must be incorrect.
$\therefore 8x + 2y \neq 3$. ∎

4. **Proof.** Let $a, b \in \mathbb{R}$.

Suppose $a > b$ and $c > 0$, $ac \leq bc$.

Then $ac - bc \leq 0$.

$\therefore c(a - b) \leq 0$.

As $c > 0$, $a - b \leq 0$.

$\therefore a < b$.

But this contradicts our assumption that $a > b$.

Therefore if $a > b$ and $c > 0$, then $ac > bc$. ∎

8. **Proof.** Let $x \in \mathbb{R}^+$ and assume that $\frac{x}{x+1} \geq \frac{x+1}{x+2}$.

Since x is positive, we know that both

$(x + 1)$ and $(x + 2)$ must be positive.

Thus, we can multiply both sides of the inequality above

by $(x + 1)(x + 2)$ to obtain $x(x + 2) \geq (x + 1)^2$.

Thus, $x^2 + 2x > x^2 + 2x + 1$.

But then , by subtracting the quantity $(x^2 + 2x)$ from both sides,

we see that $0 > 1$, which is impossible.

Hence our assumption must be incorrect, and thus $\frac{x}{x+1} < \frac{x+1}{x+2}$. ∎

2.7 Putting It All Together

11. **Proof.** Let $x, y \in \mathbb{R}$.

Suppose that it is not true that

$x \neq y$, $x > 0$, and $y > 0$ implies $\frac{x}{y} + \frac{y}{x} > 2$.

That is, suppose that $x \neq y$, $x > 0$, $y > 0$, and $\frac{x}{y} + \frac{y}{x} \leq 2$.

Then since x and y are both positive, $xy > 0$.

We can multiply both sides of the inequality $\frac{x}{y} + \frac{y}{x} \leq 2$ by xy to yield:

$x^2 + y^2 \leq 2xy$.

But then we see that $(x - y)^2 \leq 0$, and we know $(x - y) \in \mathbb{R}$.

This must mean that $(x - y)^2 = 0$.

Thus, $(x - y) = 0$ and so $x = y$.

However this contradicts our assumption that $x \neq y$.

So it is impossible that the original statement be false.

Hence, it must be true. ∎

15. **Proof.** Let $a, b, c, d \in \mathbb{R}$.

Suppose $a > b > 0$, $c > d > 0$, and $ac \leq bd$.

Since $a > b$ and $c > 0$, $ac > bc$.

Since $c > d$ and $b > 0$, $bc > dc$.

$\therefore ac > bc > dc$.

i.e. $ac > dc$. ∎

16. **Proof.** Let $x \in \mathbb{Z}$.

Note that $y = (6 - 3x) \in \mathbb{Z}$.

Furthermore, $3x + y = 3x + (6 - 3x) = 6$. ∎

20. **Proof.** Let $x \in \mathbb{Z}$.

Let $y = \frac{1-2x}{3}$. Since $x \in \mathbb{Z}$, $1 - 2x \in \mathbb{Z}$. Thus $y \in \mathbb{Q}$.

Also $2x + 3y = 2x + 3\left(\frac{1-2x}{3}\right) = 1$. ∎

22. **Proof.** Note that $\sqrt{2}$ is an irrational number and $\sqrt{2} \cdot \sqrt{2} = 2 \in \mathbb{Q}$.
    $\therefore$ the product of 2 irrationals is not necessarily an irrational number.

    Thus, the irrational numbers are not closed under multiplication. ■

2.8 Regular Induction

3. **Proof.** Note that if $n = 1$, the statement asserts
   $1^3 = [\frac{1(2)}{2}]^2$ which is true.
   So suppose $k \in \mathbb{N}$ and that $1^3 + 2^3 + ... + k^3 = [\frac{k(k+1)}{2}]^2$.
   Note that $1^3 + 2^3 + ... + k^3 + (k+1)^3 = [\frac{k(k+1)}{2}]^2 + (k+1)^3$.
   $= \frac{k^2(k+1)^2}{4} + \frac{4(k+1)^3}{4}$
   $= \frac{(k+1)^2[k^2+4(k+1)]}{4}$
   $= \frac{(k+1)^2(k^2+4k+1)}{4}$
   $= \frac{(k+1)^2(k+2)^2}{4}$
   $= [\frac{(k+1)((k+1)+1)}{2}]^2$.
   $\therefore$ BPMI, the statement is true for all $n \in \mathbb{N}$. ■

4. PP1: The proof is not correct. The author has simply assumed (not proven) that the statement is true.
   PP2: The proof contains an error. In the third sentence, the author writes, "Then $k^3 + (k+1)^3 = ...$," and the proof should state, "Then $1^3 + 2^3 + ... + k^3 + (k+1)^3 = ...$."
   If this problem were to be fixed, it would be a good proof.
   PP3: The proof contains an error in the fourth sentence. Here, the author writes, "$1^3 + 2^3 + ... + k^3 + (k+1)^3 = [\frac{(k+1)((k+1)+1)}{2}]^2$", but this is what needs to be proven.
   They must remove "$= [\frac{(k+1)((k+1)+1)}{2}]^2$" until they show this.

10. **Proof.** Note that when $n = 1$, the statement claims that
    $(1 + 5 + 6)$ is divisible by 3 which is indeed true.
    So suppose that k is a natural number for which the statement is true.
    That is, suppose we know that $k^3 + 5k + 6 = 3q$, for some $q \in \mathbb{Z}$.
    Note that $(k+1)^3 + 5(k+1) + 6 = k^3 + 3k^2 + 3k + 1 + 5k + 5 + 6$
    $= (k^3+5k+6)+(3k^2+3k+6) = (3q)+(3k^2+3k+6) = 3(q+k^2+k+2)$.
    Since $(q + k^2 + k + 2) \in \mathbb{Z}$, we see that
    $(k+1)^3 + 5(k+1) + 6$ is divisible by 3.
    Thus, the statement is true for $k + 1$.
    Therefore, BPMI, the statement must be true for all $n \in \mathbb{N}$. ■

13. **Proof.** Note that when $n = 1$ the statement claims
    that $2^0 + 2^1 = 2^{(1+1)} - 1$, i.e. $1 + 2 = 4 - 1$.
    Hence, the statement is true when $n = 1$.
    Now suppose that $k \in \mathbb{N}$ and that $2^0 + 2^1 + \cdots + 2^k = 2^{(k+1)} - 1$.
    Note that $2^0 + 2^1 + \cdots + 2^k + 2^{(k+1)} = (2^0 + 2^1 + \cdots + 2^k) + 2^{(k+1)}$
    $= (2^{(k+1)}) - 1) + 2^{(k+1)} = 2 \cdot 2^{(k+1)} - 1 = 2^{(k+2)} - 1 = 2^{((k+1)+1)} - 1$.
    Thus, BPMI, the statement is true for all $n \in \mathbb{N}$. ■

2.9 Induction with Inequalities

3. **Proof.** Note that when $n = 3$, the statement asserts that $3^2 > 6 + 1$,
   which is indeed true.
   Hence the statement is true when $n = 3$.
   Next suppose that $k \in \mathbb{N}$, $k \geq 3$, and that $k^2 > 2k + 1$.
   Note then that $(k + 1)^2 = k^2 + 2k + 1 > (2k + 1) + 2k + 1$
   $= (2k + 3) + (2k - 1)$.
   And since we know $k \geq 3$, we certainly know that $(2k - 1) > 0$.
   Thus, $(2k + 3) + (2k - 1) > (2k + 3) = 2(k + 1) + 1$.
   Hence we have shown that $(k + 1)^2 > 2(k + 1) + 1$.
   So the statement is true when $n = k + 1$.
   Thus, BPMI, the statement is true for all $n \in \mathbb{N}$, where $n \geq 3$. ∎

8. **Proof.** Note that when $n = 5$ the statement claims
   that $6! > 2^8$. i.e. $720 > 526$.
   Hence, the statement is true when $n = 5$.
   So suppose that $k \in \mathbb{N}$, $k \geq 5$, and that $(k + 1)! > 2^{k+3}$.
   Note that $((k + 1) + 1)! = (k + 2)(k + 1)! > (k + 2)2^{k+3}$.
   Now since $k \geq 5$, we certainly know that $(k + 2) > 2$.
   Furthermore, since we know 2^{k+3} is positive,
   we know that $(k + 2) \cdot 2^{k+3} > 2 \cdot 2^{k+3}$.
   Furthermore, $2 \cdot 2^{k+3} = 2^{k+4} = 2^{((k+1)+3)}$
   Thus, we see that $((k + 1) + 1)! > 2^{((k+1)+3)}$.
   Hence, the statement is true when $n = k + 1$.
   BPMI, the statement must be true for all $n \in \mathbb{N}$ with $n \geq 5$. ∎

2.10 Recursion and Extended Induction

3. **Proof.** Note that when $n = 1$ the formula yields $a_1 = 3 - 1 = 2$.
   ∴ the statement is true when $n = 1$.
   Also note that when $n = 2$, the formula yields $a_2 = 3 - 2 = 1$.
   So the statement is also true when $n = 2$.
   Now suppose that $k \in \mathbb{N}$, $k \geq 2$, and the formula
   holds true for a_k and a_{k-1}.
   That is, suppose that $k \in \mathbb{N}$, $k \geq 2$, $a_k = 3 - k$, and $a_{k-1} = 3 - (k - 1)$.
   Note that as $k + 1 \geq 3$, $a_{k+1} = 2a_k - a_{k-1}$.
   And, from our inductive assumption, $2a_k - a_{k-1}$
   $= 2(3 - k) - (3 - (k - 1))$
   $= 6 - 2k - 3 + k - 1$
   $= 2 - k = 3 - (k + 1)$.
   ∴ the statement is true when $n = k + 1$.
   BPMI, it must be true for all $n \in \mathbb{N}$. ∎

8. **Proof.** Note that when $n = 0$, the statement asserts that $h_0 \leq 3^0$,
i.e. $h_0 \leq 1$.
Thus, the statement is true when $n = 0$.
Also note that when $n = 1$, the statement asserts that $h_1 \leq 3^1$.
So, we see the statement is also true when $n = 1$.
Finally, note that when $n = 2$, the statement asserts that $h_2 \leq 3^2$.
Again, we see the statement holds true.
Now suppose that $k \in \mathbb{N}$, $k \geq 2$, and the statement
is true for h_k, h_{k-1}, and h_{k-2}.
That is, suppose that $h_k < 3^k$, $h_{k-1} < 3^{(k-1)}$, and $h_{k-2} < 3^{(k-2)}$.
As $(k+1) \geq 3$, $h_{k+1} = h_k + h_{k-1} + h_{k-2}$.
By applying our inductive assumption we see that
$h_{k+1} < 3^k + 3^{k-1} + 3^{k-2}$.
Furthermore, $3^k + 3^{k-1} + 3^{k-2} = 3^{k-2}(3^2 + 3 + 1)$
$= 3^{k-2}(13) < 3^{k-2}(27) = 3^{k-2} \cdot 3^3 = 3^{(k+1)}$.
Therefore the statement is true for $n = k + 1$.
BPMI, it must be true for all $n \in \mathbb{Z}$, with $n \geq 0$. ∎

10. **Proof.** Note that when $n = 1$ the statement asserts that $f_1^2 = f_1 \cdot f_2$.
This is indeed true as $1^2 = 1 \cdot 1$.
Note also that when $n = 2$ the statement asserts that $f_1^2 + f_2^2 = f_2 \cdot f_3$.
This is also true since $1^2 + 1^2 = 1 \cdot 2$.
So suppose that $k \in \mathbb{N}$, $k \geq 2$, and that the statement is true for k.
That is, suppose we know that $f_1^2 + f_2^2 + \dots + f_k^2 = f_k \cdot f_{k+1}$.
Note that $f_1^2 + f_2^2 + \dots + f_k^2 + f_{k+1}^2 = (f_k \cdot f_{k+1}) + f_{k+1}^2 = f_{k+1}(f_k + f_{k+1})$.
Now as $(k+2) \geq 3$, we know that $f_{k+2} = f_{k+1} + f_k$.
Thus, we see that $f_{k+1}(f_k + f_{k+1}) = f_{k+1} \cdot f_{k+2}$.
That is, $f_1^2 + f_2^2 + \dots + f_k^2 + f_{k+1}^2 = f_{k+1} \cdot f_{k+2}$.
$\therefore$ the statement is true for $n = k + 1$.
BPMI, it must be true for all $n \in \mathbb{N}$. ∎

2.11 Uniqueness Proofs, the WOP, and a Proof of the Division Algorithm

4. **Proof.** Note that $\frac{-4}{3} \in \mathbb{R}$, and if we let $x = \frac{-4}{3}$
then $f(x) = 3\left(\frac{-4}{3}\right) + 1 = -3$, and $g(x) = 6\left(\frac{-4}{3}\right) + 5 = -8 + 5 = -3$.

Next suppose that $x_1, x_2 \in \mathbb{R} \ni f(x_1) = g(x_1)$ and $f(x_2) = g(x_2)$.
Then $3x_1 + 1 = 6x_1 + 5$ and $3x_2 + 1 = 6x_2 + 5$.
Combining terms in each of these equations yields
$-4 = 3x_1$ and $-4 = 3x_2$.
Thus, $3x_1 = 3x_2$.
$\therefore x_1 = x_2$. ∎

7. **Proof.** Let $x \in \mathbb{Z}$.
Note that $y = -x \in \mathbb{Z}$ and $x + y = x + -x = 0$.
To see that y is unique, suppose $y_1 \in \mathbb{Z}$ and $y_2 \in \mathbb{Z}$,
where $x + y_1 = 0 = x + y_2$.
Since $x + y_1 = x + y_2$, $y_1 = y_2$.
$\therefore y$ is unique. ∎

8. This set has no least element.

15. This set does not have a least element.

16. This set has a least element, it is 1.

Chapter 3

3.1 Sets

6. $\subseteq$

7. neither

23. $\{\varnothing, \{\varnothing\}, \{\{1\}\}, \wp(\{1\})\}$

3.2 Set Operations

3. $\{0, 1, 2, 3, 4, 5, 6, 7, 8\}$

7. $\{3, 1, \{1\}, 2, \{3, 2\}, \{\varnothing\}\}$

10. A

3.3 Set Theory

4. Note that if $A = \{1, 2\}, B = \{8\}$, and $C = \{1, 3\}$, then $A - (B - C) = \{1, 2\}$ and $(A - B) - C = \{2\}$. Here $A \nsubseteq \{2\}$.

7. **Proof.** Let A, B, C and D be sets and suppose that $A \cup B \subseteq C \cup D$, $A \cap B = \varnothing$, and $C \subseteq A$.
   Suppose $x \in B$.
   Since $x \in B$, $x \in A \cup B$.
   Also, since $A \cup B \subseteq C \cup D, x \in C \cup D$.
   Thus $x \in C$ or $x \in D$.
   If $x \in C$, then as $C \subseteq A$, we know that $x \in A$.
   But this would mean that $x \in A \cap B$,
   and this is not possible since $A \cap B = \varnothing$.
   Thus it must be the case that $x \in D$.
   $\therefore B \subseteq D$. ■

14. **Proof.** Let A, B, C and D be sets.
    Suppose $A \subseteq B, C \subseteq D$, and $B \cap D = \varnothing$.
    By way of contradiction suppose that $\exists x \ni x \in A$ and $x \in C$.
    As $x \in A$, and $A \subseteq B$, then $x \in B$.
    Also, as $x \in C$, and $C \subseteq D$, then $x \in D$.
    But, this means that $x \in B \cap D$ which is impossible since $B \cap D = \varnothing$.
    Therefore, A and C are disjoint. ■

3.4 Indexed Families of Sets

3. $\mathbb{N}$

7. $[2, 5]$

11. **Proof.** Suppose $x \in \bigcup_{\alpha \in \Lambda} A_\alpha$ and $x \in \bigcap_{\alpha \in \Lambda} A_\alpha^c$.

As $x \in \bigcap_{\alpha \in \Lambda} A_\alpha^c$, $\forall \alpha \in \Lambda$, $x \in A_\alpha^c$.

$\therefore \forall \alpha \in \Lambda$, $x \notin A_\alpha$.

But if x is not in any of the A_α's then it is impossible

that $\exists \alpha \in \Lambda \ni x \in A_\alpha$.

That is, it is not true that $x \in \bigcup_{\alpha \in \Lambda} A_\alpha$.

$\therefore$ the sets are disjoint. $\blacksquare$

3.5 Cartesian Products

4. $\{(a, 0), (a, 1), (a, 2), (b, 0), (b, 1), (b, 2), (c, 0), (c, 1), (c, 2)\}$

10. **Proof.** Let A and B be non empty sets.
    By contrapositive, suppose that $A = B$.
    Let $(x, y) \in A \times B$.
    Then $x \in A$ and $y \in B$.
    Since $A = B$, we know that $x \in B$ and $y \in A$.
    $\therefore (x, y) \in B \times A$.
    Similarly, if $(x, y) \in B \times A$ then $(x, y) \in A \times B$.
    $\therefore A \times B = B \times A$. $\blacksquare$

14. **Disproof.** Note when $B = \{1\}, C = \{1\}$, and $A = \{3\}$.
    Hence $(A \cap B) \times C \neq (A \times B) \cap (A \times C)$. $\blacksquare$

3.6 Relations

3. There is a problem since the definition of "approximately equal to" is not
    clear. It is quite possible that x is within an acceptable margin of error
    from y and y is an acceptable margin of error from z but that x and z are
    not close enough to be called "approximately equal."

10. **Proof.** Let $x \in \mathbb{Z} - \{0\}$.
    Note that as $x \neq 0, x \cdot x = x^2 > 0$.
    $\therefore x \sim x$, and so $\sim$ is reflexive.
    Next suppose that $x \sim y$.
    Then $x \cdot y > 0$, and since $x \cdot y = y \cdot x$, we know that $y \cdot x > 0$.
    Thus $\sim$ is symmetric.
    Finally, suppose that $x \sim y$ and $y \sim z$.

Then $x \cdot y > 0$ and $y \cdot z > 0$.
$\therefore x$ and y are either both positive or both negative,
and y and z must also have the same sign.
Thus, x and z must have the same sign, $\therefore x \cdot z > 0$.
Hence, $\sim$ is transitive.
$\therefore \sim$ is an equivalence relation. ∎

[1] and [−1] are the only (distinct) equivalence classes formed by this relation.

13. **Proof.** Let $(a, b) \in \mathbb{R} \times (\mathbb{R} - \{0\})$.
Note that since $ab = ab$, $(a, b)R(a, b)$.
$\therefore R$ is reflexive.
Next suppose that $(a, b)R(c, d)$.
Then $ad = bc$.
$\therefore cb = da$.
Thus, $(c, d)R(a, b)$, and so R is symmetric.
Finally, suppose that $(a, b)R(c, d)$ and $(c, d)R(e, f)$.
Then $ad = bc$ and $cf = de$.
As $d \neq 0$, $a = \frac{bc}{d}$.
$\therefore af = \frac{bc}{d}f = b\frac{cf}{d} = b\frac{de}{d} = be$.
Thus, $(a, b)R(e, f)$, and so R is transitive.
$\therefore R$ forms an equivalence relation. ∎

3.7 Functions

3. f is not a function since $f(\frac{1}{3}) \notin \mathbb{Z}$.

10. **Proof.** Let $z \in \mathbb{Z}$. Note that $(z, 3) \in \mathbb{Z} \times \mathbb{Z}$ and $f((z, 3)) = z - 3 + 3 = z$.
$\therefore f$ is onto. ∎

14. **Proof.** Suppose $f(x) = f(y)$.
Then $(2x, x^2) = (2y, y^2)$.
$\therefore 2x = 2y$ and $x^2 = y^2$.
Since $2x = 2y$, $x = y$, $\therefore f$ is $1 - 1$.
To see that f is not onto, note that $(1, -1) \in \mathbb{Z} \times \mathbb{Z}$
and $(1, -1)$ has no preimage.
To see this, suppose $\exists a \in \mathbb{Z} \ni f(a) = (1, -1)$.
Then $2a = 1$ and $a^2 = -1$.
But there does not exist $\mathbb{Z} \ni a^2 = -1$.
$\therefore f$ is not onto. ∎

19. **Proof.** Let $q \in \mathbb{Q}$.
Then $\exists a, b \in \mathbb{Z}, b \neq 0 \ni q = \frac{a}{b}$.
Note that $(a + 1, b) \in \mathbb{Z} \times \mathbb{N}$ and $f((a + 1, b)) = \frac{a+1-1}{b} = q$.
$\therefore f$ is onto.
To see that f is not $1-1$, note that $(11, 10) \neq (3, 2)$ but $f((11, 10)) = 1$
$= f((3, 2))$.
$\therefore f$ is not $1 - 1$. ∎

20. (a) $\{(0, n) | n \in \mathbb{N}\}$
    (b) $\{(3, 1), (6, 2), (9, 3), ...\} = \{(3n, n) | n \in \mathbb{N}\}$

3.8 Composition of Functions

4. **Proof.** Note that $1 \in \mathbb{R}$ which has no preimage under $g \circ f$.
To see this, suppose $\exists x \in \mathbb{Z} \ni (g \circ f)(x) = 1$.
Then $g(f(x)) = 1$.
$\therefore g(\frac{x}{2}) = 1$.
$\therefore \frac{\pi x}{2} = 1$.
$\therefore \pi x = 2$.
$\therefore x = \frac{2}{\pi} \notin \mathbb{Z}$.
$\therefore 1$ has no preimage under $g \circ f$.
$\therefore g \circ f$ is not onto. $\blacksquare$

3.9 Cardinality

5. **Proof.** Let $E = \{x \in \mathbb{Z} | \exists k \in \mathbb{Z} \ni x = 2k\}$.

Define $f : \mathbb{N} \to E$ by $f(n) = \begin{cases} n-1 & \text{if} \quad n \text{ is odd} \\ -n & \text{if} \quad n \text{ is even} \end{cases}$.

Suppose $f(n_1) = f(n_2)$.
Then one of the following must be true:
$n_1 - 1 = n_2 - 1, \ -n_1 = -n_2, \ n_1 - 1 = -n_2,$ or $n_1 - 1 = -n_2.$
The last two cannot be true as either of these would imply
that a negative number equals a non-negative number.
Hence, one of the first two must be true.
But note that both of the first two imply that $n_1 = n_2$.
$\therefore f$ is $1 - 1$.
Next suppose $e \in E$.
Either $e \geq 0$ or $e < 0$.
If $e \geq 0$, then $e + 1 \in \mathbb{N}$.
Further since $e + 1$ is odd, $f(e + 1) = (e + 1) - 1 = e$.
If $e < 0$, then $-e > 0$ and so $-e \in \mathbb{N}$.
Further as e is even, $f(-e) = -(-e) = e$.
$\therefore f$ is onto.
$\therefore \exists$ a bijection $f : \mathbb{N} \to E$.
$\therefore E$ is countably infinite. $\blacksquare$

10. **Proof.** Let $f : (0, 2) \to (-10, 10)$ by $f(x) = 10x - 10$.
Note that f is $1 - 1$ for if $f(x_1) = f(x_2)$, then $10x_1 - 10 = 10x_2 - 10$.
And so, $10x_1 = 10x_2$.
$\therefore x_1 = x_2$.
Furthermore, if $y \in (-10, 10)$, then $\frac{y+10}{10} \in (0, 2)$.
And $f(\frac{y+10}{10}) = 10(\frac{y+10}{10}) - 10 = y + 10 - 10 = y$.
$\therefore f$ is onto.
$\therefore f$ is a bijection.
$\therefore (0, 2)$ has the same cardinality as $(-10, 10)$. $\blacksquare$

13. **Proof.** Define a relation $\sim$ on the set of all sets by
$A \sim B$ iff A and B have the same cardinality.
Certainly $\sim$ is reflexive because we proved in Problem 12 that
any set has the same cardinality as itself.
Now, suppose A and B are sets $\ni A \sim B$.
Then $\exists$ a $1-1$ and onto function $f : A \to B$.
As f is $1-1$ and onto, f^{-1} exists and is a $1-1$ and
onto function from $B \to A$.
$\therefore B$ has the same cardinality as A.
$\therefore B \sim A$.
$\therefore \sim$ is symmetric.
Finally, suppose that A, B, and C are sets $\ni A \sim B$ and $B \sim C$.
Then $\exists 1-1$ and onto functions g and $f \ni g : A \to B$ and $f : B \to C$.
Note that we proved in the homework of Section 3.8 that if
$g : A \to B$ and $f : B \to C$ are both $1-1$ and onto functions,
then $f \circ g : A \to C$ is a $1-1$ and onto function.
$\therefore A \sim C$.
$\therefore \sim$ is an equivalence relation. ■

Index

Commonly Used Mathematical Notations

$\mathbb{N}$	The set of natural numbers
$\mathbb{Q}$	The set of rational numbers
$\mathbb{R}$	The set of real numbers
$\mathbb{Z}$	The set of integers
$\mathbb{Z}^+$	The set of positive integers
$\mathbb{Z}^*$	The set of non-negative integers
$\forall$	For all
$\exists$	There exists
$\exists!$	There exists a unique
$\in$	is an element of
$\ni$	such that
$\therefore$	Therefore
iff	if and only if
$a\mid b$	a divides b
$a \equiv b \bmod n$	a is congruent to b modulo n
$\varnothing$	The empty set
$A \cup B$	The union of A and B
$A \cap B$	The intersection of A and B
$A \subseteq B$	A is a subset of B
$A \times B$	The Cartesian product of A and B
$\wp(A)$	The power set of A
A^c	A complement